Meeting at Grand Central

UNDERSTANDING THE SOCIAL AND
EVOLUTIONARY ROOTS OF COOPERATION

Lee Cronk and Beth L. Leech

PRINCETON UNIVERSITY PRESS

Princeton and Oxford

Copyright © 2013 by Princeton University Press
Published by Princeton University Press,
41 William Street, Princeton, New Jersey 08540
In the United Kingdom: Princeton University Press,
6 Oxford Street, Woodstock, Oxfordshire OX20 1TW

press.princeton.edu

ISBN 978-0-691-15495-4

Library of Congress Cataloging-in-Publication Data

Cronk, Lee.
    Meeting at Grand Central : understanding the social and evolutionary
roots of cooperation / Lee Cronk and Beth L. Leech.
        p. cm.
    Includes bibliographical references and i ndex.
    ISBN 978-0-691-15495-4 (alk. paper)
    1. Cooperation—History.   2. Social interaction—History.   I. Leech, Beth L.,
1961– II. Title.
    HD2956.C76 2012
    303.3—dc23   2012015008

British Library Cataloging-in-Publication Data is available

This book has been composed in Sabon LT

Printed on acid-free paper. ∞

Printed in the United States of America

10 9 8 7 6 5 4 3 2 1

Meeting at Grand Central

For Judy, who didn't believe in the collective action dilemma

Contents

Preface   ix

Chapter 1   Cooperation, Coordination, and Collective Action   1
    Box 1.1
    EXPERIMENTAL ECONOMIC GAMES   15

Chapter 2   Adaptation: A Special and Onerous Concept   18

Chapter 3   The Logic of *Logic*, and Beyond   47
    Box 3.1
    TYPES OF GROUPS   49
    Box 3.2
    TYPES OF GOODS   53

Chapter 4   Cooperation and the Individual   72
    Box 4.1
    THE RECIPROCITY BANDWAGON   75
    Box 4.2
    THE PRISONER'S DILEMMA GAME   79

Chapter 5   Cooperation and Organizations   101

Chapter 6   Meeting at Penn Station:
Coordination Problems and Cooperation   124
    Box 6.1
    COORDINATION GAMES   150

Chapter 7   Cooperation Emergent   151

Chapter 8   Meeting at Grand Central   169

Notes   189

References   207

Index   237

Preface

WE TEACH ON RUTGERS UNIVERSITY'S CAMPUS in New Brunswick, New Jersey, located in the southern part of the New York metropolitan area. On September 11, 2001, Beth was scheduled to teach a seminar on citizen activism. The class did not meet that day. Following that morning's terrorist attacks, the university cancelled all classes. When classes resumed the following week, the syllabus called for a discussion of Mancur Olson's writings on the collective action dilemma. Olson, an economist, used a formal model to explain why "rational self-interested individuals will not act to achieve their common or group interests."[1] The problem, Olson argued, is that a rational person would instead choose to let someone else do the work while still reaping the benefits. Because this is a problem that every group must face, the class began with the issue of what to do when most people "free ride," or choose not to contribute to the common good.

The students were astounded by the very question. "What do you mean?" they asked. "OF COURSE people participate. EVERYONE participates. Just look at what people are doing now in response to the terrorist attacks." And they had a point. Thousands of people volunteered at Ground Zero, where the World Trade Center's twin towers once stood. Many thousands more around the world responded to calls for donations of money and relief supplies, with cash donations by individuals totaling more than $1.5 billion.[2] That generosity was accompanied, particularly in New York City itself, by a refreshing spirit of civility, friendliness, and cooperativeness that lasted for some time after the initial shock of the attacks had subsided.

We are thus faced with a paradox: theory predicts that cooperation will be rare, but everyday experience tells us that it is quite common. This paradox is what drew us to the study of cooperation and inspired us to write this book. Most scholarship to date has focused on only one side or the other of this paradox. Which side any particular scholar focuses on depends largely upon which body of scholarly literature most informs his or her worldview. Social scientists, particularly political scientists, economists, and sociologists, tend to focus on free riding and other obstacles to collective action and how they are sometimes overcome. Scholars with an interest in human evolution, in contrast, tend to be most impressed by how much humans cooperate, particularly with nonrelatives, compared to most other species. We represent both approaches. Beth L. Leech is a political scientist whose work on such topics as interest groups, agenda setting, and lobbying has been grounded in social science theories of policy formation and group formation.[3] Lee Cronk is an anthropologist whose work on such topics as social norms and cultural change has been grounded in both evolutionary theory and an appreciation of culture's influence on behavior.[4]

Our view is that this paradox is more apparent than real. A full understanding of human cooperation requires an appreciation of both sides of the coin. Although it is true that free riding and other problems often stand in the way of successful collective action, it is also true that people have found a wide variety of ways to overcome and circumvent those problems. Although it is true that humans cooperate with each other much more than do members of most other species, it is also true that humans do a wide range of things much more than any other species: play chess, drive cars, drink coffee, use language, create art, and so on. Our abilities to use language and create art may be the result of selection for those characteristics, but our abilities to drive cars and drink coffee are clearly just by-products of selection for other characteristics. If our high levels of cooperation are the result of selection in favor of characteristics that make cooperation more likely, then we need theories about how those selection pressures would have been felt among our ancestors. However, if some of our species' success with cooperation is, like playing chess and driving cars, not the product of selection for cooperation specifically but rather the outgrowth of other characteristics that were favored by selection, then we may not need special theories to explain it. Reality is probably some mix of these two extreme possibilities.

Furthermore, comparing human cooperation levels with those found in other species is not necessarily the most appropriate way to put human cooperation in perspective. A better comparison might be between the real and the ideal: What proportion of situations in which people could benefit from cooperating with each other actually result in cooperation? Despite how much we cooperate, it may be that we actually do so far less often than we could if only we were better at overcoming the collective action dilemma or coordinating our social behaviors. Claims that cooperation is either rare or common inevitably come up against a kind of epistemological roadblock: We know only about the instances in which people at least tried to cooperate. We know nothing about the instances in which people might have benefited from cooperation but were unaware of either the potential to benefit or the best way to go about it. The deck is stacked in favor of the conclusion that humans are great cooperators.

In the chapters that follow, we address these issues and many others related to human cooperation and how it is studied. Our hope is to foster a common culture—shared terminologies, concepts, and theoretical insights—among the diverse range of scholars who study cooperation and their students.

The first step we took toward this book came in the spring of 2007, when we cotaught an interdisciplinary graduate seminar on cooperation and collective action. We would like to thank the students in that seminar for our many stimulating discussions. During the 2008–9 academic year, we were on leave in Princeton, New Jersey, Lee at the Institute for Advanced Study and Beth at both IAS and the Center for the Study of Democratic Politics at Princeton University's Woodrow Wilson School of Public and International Affairs. We would like to thank both institutions for their generous support and for creating an atmosphere that was both stimulating and contemplative. We would especially like to thank our host at IAS, Eric Maskin, and Beth's host at CSDP, Larry Bartels, who coincidentally both got us thinking about how strategic voting is a kind of cooperation. We would also like to acknowledge Rutgers University's willingness to allow us to take the year off teaching so that we could take advantage of those opportunities. In Lee's case, Rutgers contributed some additional funding to make this possible.

C. Athena Aktipis, Frank Batiste, Frank Baumgartner, Derek Bickerton, Rolando de Aguiar, D. Bruce Dickson, Bria Dunham, Drew Gerkey, William Irons, Padmini Iyer, Dominic D. P. Johnson, Jack Levy, Robert Lynch, Doug Pierce, Eric Radezky, Montserrat Soler, Helen Wasielewski, and an anonymous reviewer read full or partial drafts of our manuscript, and the comments they provided helped us a great deal in refining our arguments and presentation. Helpful feedback was also provided by participants in the Rutgers political science department's Emerging Trends speaker series, by students in Lee's course on evolution and cooperation and Beth's graduate course on organized interests. Finally, we would like to thank Chuck Myers at Princeton University Press for his encouragement and the expert way he shepherded our book through the publication process. Of course, we retain responsibility for any errors or shortcomings that remain.

Meeting at Grand Central

CHAPTER 1

Cooperation, Coordination, and Collective Action

• •

Fresh water is a scarce resource around the world, but particularly in arid regions such as the American West. At one time, groundwater was sufficient for the needs of the region's small population, but rapidly growing populations in recent years have led to the depletion of aquifers and the diversion of enormous amounts of water from the Colorado and other rivers. Conservation efforts have been, for the most part, sporadic and ineffective, and, for many communities in the region, a water crisis looms on the horizon.[1] In contrast, communities of farmers around the world have been successfully sharing irrigation water for many years. In some cases, such arrangements have existed for centuries. Farmers in Valencia, Spain, for example, still use rules for water distribution that were drawn up in 1435.[2]

Citizens of many countries are frustrated by ineffective and corrupt police forces. Although they may occasionally take matters into their own hands on an individual and ad hoc basis, it is rare for them to organize a viable alternative to the police. In Tanzania in the 1980s, such a rare event did occur. Members of the Sukuma ethnic group organized Sungusungu, a system of grassroots justice and vigilantism. Sungusungu was so successful that it was deputized by the Tanzanian government and imitated by other Tanzanian ethnic groups. But the imitators did not always share the success of the Sukuma. Members of the Pimbwe ethnic group, for example, attempted to form their own Sungusungu, but they eventually abandoned the effort in frustration.[3]

Slavery has existed in a wide variety of societies throughout history, but the Atlantic slave trade was by far the largest. Between the fifteenth and nineteenth centuries, approximately ten million Africans were captured and shipped to the Americas, and millions of their descendants lived their entire lives as slaves.[4] Despite the fact that slaves outnumbered slave owners in many areas, slave rebellions did not occur often, and they were seldom successful. One of the rare exceptions occurred on board *La Amistad*, a slave ship, in 1839. After the ship docked in Connecticut, U.S. courts eventually found that the would-be slaves were free, and they returned to Africa.[5]

Every day, tens of thousands of commercial aircraft take off and land around the world. Many airports are extraordinarily busy, with some handling thousands of flights a day. Despite all of that traffic in the air and on the ground, collisions between planes are very rare. One such rare and tragic event resulted in the worst air disaster in history. On March 27, 1977, on the Spanish island of Tenerife, a Boeing 747 owned by KLM collided on takeoff with a Pan Am 747 taxiing on the ground. All of the 248 people aboard the KLM plane died, as did 335 people on the Pan Am jet.[6]

....

An eclectic topic needs an eclectic approach

Water management, vigilante movements, slave rebellions, and colliding aircraft. What could phenomena this diverse possibly have in common? They are all examples of cooperation of one kind or another, contrasted in each case with a similar situation in which cooperation failed to occur. We chose to begin our book with these examples to drive home an important point: Cooperation may occur all around us every day, but it should not be taken for granted. The forces working against successful cooperation are formidable, and the fact that cooperation occurs as often as it does is remarkable and noteworthy.

That, in a nutshell, is why we wrote this book.

In our final chapter, we will return to those four examples. For now, they serve to demonstrate that cooperation—working together to achieve a common goal—is a very broad and diverse phenomenon that includes a wide range of specific behaviors. Writing this book together is an act of cooperation. As we go about the rest of our daily lives, we cooperate in a wide variety of ways and with a wide range of people. At work, we coordinate our actions with coworkers and students in order to get our classes taught, our exams graded, our graduate students funded, and so on. In our community, we cooperate with other citizens when we volunteer with local nonprofits, attend events at our children's schools, participate in civic organizations, and vote. When we shop, we cooperate with storekeepers through mutually beneficial exchanges of money for goods and services. When we drive, we cooperate with pedestrians, bicycle riders, and other drivers so that we may all get where we want to go. When we play, we cooperate not only with our teammates but also with members of the other team by adhering to a set of mutually agreed-upon rules.

Because cooperation is such a large and diverse phenomenon, understanding it will require a large tool kit of ideas drawn from a wide range of disciplines. In philosophers' jargon, the study of cooperation needs to be a "historical" science rather than a "theoretical" or "Newtonian" one.[7] The theoretical sciences are the simple, elegant ones that produce general theories of narrowly defined phenomena. Their simplicity and elegance is an outgrowth of how they define their subject matters. Physics, for example, has beautiful formal theories of the phenomena it studies because it studies only a very limited and carefully defined range of phenomena. When things get beyond that range, they become the realm of either another theoretical science (e.g., chemistry), a historical science (e.g., cosmology), or engineering. In the life sciences, the broad field of evolutionary biology includes a very well developed theoretical science regarding how evolution works in the abstract. In the social sciences, economics is the best developed theoretical science. As with physics, economics' ability to develop in that way is a result of the narrow way that economists define their subject matter.

Historical sciences, in contrast, make use of whatever they need—insights from theoretical sciences, empirical observations, etc.—in order to explain their complex, varied, and roughly bounded subject matters. Historical sciences are, in a word, eclectic.[8] In the physical sciences, they include such fields as geology, astronomy, and meteorology. In the life sciences, they include paleontology and animal behavior studies. In the social sciences, they include anthropology, sociology, geography, most of political science, and some aspects of economics. To illustrate the way historical sciences approach their phenomena, economist Friedrich Hayek suggested that we imagine what a scientist might do if her task were to study how a garden fills up with weeds. She would need to record a large number and wide variety of details about the garden, including its soil types, patterns of shade and sunlight, plant species, and so on. To understand her data, she would need to incorporate theoretical and empirical insights from a variety of fields, including chemistry, geology, and biology.[9] Historian of science David B. Kitts has pointed out that paleontology, a biological science, must make use of the geological concept of superposition (i.e., newer material is usually on top of older material) in order to do one of its most important jobs, that of determining fossils' relative ages.[10] Similarly, absolute dating techniques, such as radiocarbon and argon-argon, require physics. For a historical science, such eclecticism is a source of strength. We will approach the study of cooperation in this same eclectic manner. This means that although we will certainly make use of existing formal theories regarding specific types of cooperation, we will not attempt to create a formal, mathematical theory that single-handedly explains all human cooperation. Rather, we will be discussing a wide range of ideas, theories, and existing empirical research relevant to the study of the complex and diverse phenomenon of human cooperation.

Olson, Williams, . . .

We will be drawing in particular upon two bodies of work that began separately but nearly simultaneously with the publication of Mancur Olson's *The Logic of Collective Action* in 1965 and George C. Williams's *Adaptation and Natural Selection* in 1966.[11] Although Olson was an economist and Williams an evolutionary biologist, they dealt with similar issues and presented similar arguments. Both were arguing against scholars in their fields who had emphasized cooperation's group-level benefits and discounted its individual-level costs. Both explained why a focus on groups would not provide a complete understanding of collective action and other social behaviors.

Mancur Olson argued that unless groups are small or "there is coercion or some other special device to make individuals act in their common interest, rational, self-interested individuals will not act to achieve their common or group interest."[12] Olson's challenge led to work by social scientists regarding the obstacles people face when they might benefit from collective action and how even "rational self-interested individuals" can find ways to overcome them. As the civil rights movement and other mass mobilizations have shown empirically, efforts to overcome collective action dilemmas can be helped significantly by existing social institutions and networks. In some circumstances, entrepreneurial individuals will agree to pay more than their fair share of the costs of solving a collective action problem in hopes of recouping their losses in the future, sometimes in the form of salary or other remuneration from the successful group, and sometimes simply in the form of enhanced reputation.[13] Modifying the costs and benefits of participation for different kinds of participants can provide enough people with an incentive to contribute to the effort for the collective action dilemma to be overcome. Another common strategy is to break large and difficult collective action problems down into smaller, easier ones by linking small groups in networks, nested segments, and hierarchies. Many of the major ideas emerging from the Olsonian tradition have been tested and found practical application in work on how, despite the free rider problem, groups of people do sometimes successfully cooperate.

Natural selection designs organisms through the differential survival and reproduction of various kinds of self-replicators—genes, individuals, groups, and so on.[14] A major question in evolutionary theory is what happens when selection pressures at these different levels design organisms in different ways. Which level prevails? George Williams argued that, in most circumstances, natural selection at the level of individuals will be much stronger than selection at the level of groups. As a result, selection should have designed most adaptations to benefit individuals, with whatever benefits they might also provide the groups to which those individuals belong being purely fortuitous. This presented evolutionary biologists with a challenge similar to the one that Olson posed to social scientists: *If group selection is rarely a very strong force in nature, how do we explain cooperation and other prosocial behaviors?* Biologists responded to this challenge in a variety of ways. Selection at levels below the group can favor prosocial and even self-sacrificial behaviors if genetic relatives stand to benefit because the genes behind the behaviors may be passed on even if the altruistic individual fails to survive. Selection at low levels can even favor altruism toward nonrelatives if there is some chance that such kindness will eventually be repaid. More recently, intriguing arguments have been made about how our ancestors may have tipped the odds in favor of successful collective action by using physical and behavioral cues to distinguish likely cooperators from likely free riders, resulting in selection for cooperativeness and vigilance against cheaters.

. . . and many others, as well

In addition to Olson and Williams's seminal work, we will also be relying upon the contributions of a great many other scholars from a wide range of disciplines. We will try to explain their work in ways that will make sense to all our readers, even those unfamiliar with the theories and terms of a particular scholar's home discipline. The work of those many scholars comes in a variety of forms—theoretical, methodological, and empirical—and we will make use of all of them. Our emphasis will be on the complementarity we see among the varying approaches that have been taken to the study of cooperation by scholars working in different disciplines. Scientific explanations—the correct ones, anyway—are complementary to one another because they are explaining different aspects of the same universe. Just as the different bits and pieces of the universe fit together to form a coherent whole, so should different scientific theories fit together to form a coherent explanation of the universe. Evolutionary biologist Edward O. Wilson has dubbed this complementarity among the disciplines *consilience*.[15]

One common way that explanations offered by different disciplines achieve consilience with one another is by focusing on different causes of the same phenomenon. It is often helpful to separate scientific explanations of a particular phenomenon according to the causal distance on which they focus. Some explanations are focused on causes that are very immediate, or proximate, to the thing being explained. An explanation of cooperation that focuses on the motivations experienced by the individuals involved, for example, would be of this proximate variety. Many explanations of cooperation that have emerged from the social and behavioral sciences focus on such proximate causes. Given that understanding proximate causes is often crucial to solving practical problems, such as either encouraging or discouraging cooperation, this focus makes a great deal of sense. Evolutionary theory, on the other hand, takes a step away from the phenomenon in question and asks what evolutionary forces might have shaped it. Thus, a proximate explanation of cooperation in terms of motivations is complemented by an evolutionary explanation of how those motivations came to exist.

Complementarity among the disciplines that study cooperation exists at the methodological and empirical levels, as well. The methods that have been used to study cooperation are the same ones that have been used to study many other human behaviors. Experiments, observations, interviews, case studies, formal models, and agent-based models are the most prominent. All of these methods produce valuable data, but none of them is sufficient by itself. Experiments, for example, can tell us a great deal about why and how people cooperate or fail to cooperate in a controlled setting, but they are all the more valuable if they are accompanied by good qualitative descriptions of the contexts in which such cooperation arises or fails to arise in the real world. Formal and agent-based models can help us understand cooperation in abstract and refined ways, but they do not by themselves produce data about the real world.

One body of scholarship that has been extraordinarily important to the study of cooperation is game theory. Game theory is used to model situations in which an individual actor's best choice of action is dependent upon the choices that others make. Though it originated only a few decades ago, game theory is now a large, highly developed, and complex body of formal mathematics. As such, formal game theory is beyond the scope of this volume, and we have written this book in a way that should be accessible to readers who are unfamiliar with game theory. Readers who do want to familiarize themselves with formal game theory might want to start with recent books by Ken Binmore and Herb Gintis.[16] Readers interested in a more game theoretic or formal approach to the topic of cooperation in particular are directed to Russell Hardin's excellent book on collective action, Scott Ainsworth's book on interest groups, Pamela Oliver's writing on social movements, and David Barash's book on games and evolution.[17]

Game theory has also been important to the study of cooperation as a method for gathering data about real people and how they interact with each other. This method originated in economics, but it has since been borrowed by scholars in many other disciplines, including us. Experimental games come in a variety of forms, each one designed to provide insights regarding different aspects of cooperation. Box 1.1 provides an overview of some of the most common and important experimental games.

Some important definitions
Cooperation

We were motivated to write this book, in part, by some frustration we feel with how scholars have defined important terms. First among these is the term "*cooperation*" itself. We use it to refer to the very broad and diverse phenomenon of people (and other organisms) working together. The preferred British spelling, "*co-operation*,"[18] shows that this has long been the everyday meaning of the term, and it also corresponds with how it is generally used in the social sciences. Our usage contrasts with how cooperation often is defined by scholars working in the evolutionary tradition, where it is commonly equated with altruism, that is, helping another at some cost to oneself.[19] For example, Martin Nowak states that "[c]ooperation means that the donor pays a cost and the recipient gets a benefit."[20] Similarly, Natalie and Joseph Henrich declare that "[c]ooperation occurs when an individual incurs a cost to provide a benefit for another person or people."[21] Samuel Bowles and Herbert Gintis use cooperation to refer to "an individual behavior that incurs personal costs to engage in a joint activity that confers benefits exceeding these costs to other members of one's group."[22] In this same spirit, Laurent Lehmann and his coauthors give exactly the same definition for both "cooperation" and "helping": "a behavior that increases the fitness of another individual."[23] It is unclear to us why they still need both terms.[24]

The confusion among evolutionary scientists regarding the difference between cooperation and altruism seems to have two main origins. One is the term "reciprocal altruism," coined in 1971 by evolutionary theorist Robert Trivers.[25] Reciprocity is not, of course, altruistic. An act of kindness or generosity that is repaid is, by definition, not altruistic because, at the end of the exchange, the actor experiences a benefit, not a cost. An act of kindness or generosity that is not repaid, in contrast, is not reciprocal and would be selected against. Thus, the term "reciprocal altruism," taken literally, describes a phenomenon that natural selection would never favor.[26] Despite these shortcomings, the term "reciprocal altruism" was attractive for two reasons. First, it emphasized the risk that one takes when being generous to another with no guarantee that one's generosity would ever be repaid. This helped clarify the theoretical problem that such acts present: why would selection *ever* favor an organism that incurs a cost to help another? Second, the term "reciprocal altruism" linked the study of reciprocity to the study of altruism, a topic that had generated revolutionary new insights into how natural selection works.[27] One unfortunate and unintended consequence of the term, however, has been decades of confusion among evolutionary biologists regarding the distinctions among cooperation, reciprocity, and altruism.

Confusion regarding the difference between cooperation and altruism is also the result of the domination of the evolutionary biological literature on cooperation by a simple conceptual game called the Prisoner's Dilemma. In the Prisoner's Dilemma, which we will discuss more fully in later chapters, two players are presented with a choice of two strategies, usually labeled "cooperate" and "defect." The "cooperate" strategy involves foregoing a higher payoff yielded by the "defect" strategy. Thus, "cooperating" in the Prisoner's Dilemma game is altruistic, at least in the short run.[28] Given how many scholars have relied upon the Prisoner's Dilemma to study cooperation, it is not surprising that so many have simply equated the altruistic "cooperate" strategy within the game with cooperation in general. As useful as the Prisoner's Dilemma has been to the study of cooperation, however, it is by no means the only game in town. We see no reason to equate the broad, diverse, real-world phenomenon of cooperation with a strategy in a single, narrowly defined type of game.

Conflating cooperation and altruism also creates confusion regarding what it means to explain cooperation. Some studies that purport to be about cooperation are really about people's cooperativeness, that is, their willingness to engage in cooperation, their agreeableness, their trustworthiness, their willingness to trust others, their willingness to behave altruistically toward others, or some other characteristic of individuals. Although such characteristics may help foster cooperation in many circumstances, they should not be equated with cooperation. While cooperativeness, prosociality, and so on are characteristics of individuals, cooperation is a behavior involving at least two individuals. Indeed, cooperativeness, prosociality, and a tendency to be altruistic may often be an insufficient precondition for cooperation to actually occur. A more important factor may be our ability to coordinate our behaviors with those of others.

Collective action and the collective action dilemma

People were aware of the collective action dilemma long before it had a name. Many of us are first exposed to it in school, not because it is explicitly taught, but because we are required to do group projects. Although all members of the group would benefit from a job well done, everyone has an incentive to let others do the work. Often, some do all or most of the work while others—referred to as "free riders"—do little or none but still get the same grade as those who contributed. Although much modern research on the collective action dilemma began with Olson, the idea has been part of scholarly discourse for much longer. One early description of it comes from nineteenth-century political economist William Forster Lloyd:

> Suppose the case of two persons agreeing to labour jointly, and that the result of their labour is to be common property. Then, were either of them, at any time, to increase his exertions beyond their previous amount, only half of the resulting benefit would fall to his share; were he to relax them, he would bear only half the loss. If, therefore, we may estimate the motives for exertion by the magnitude of the personal consequences expected by each individual, these motives would in this case have only half the force, which they would have, were each labouring separately for his own individual benefit. Similarly, in the case of three partners, they would have only one third of the force—in the case of four, only one fourth—and in a multitude, no force whatever. For beyond a certain point of minuteness, the interest would be so small as to elude perception, and would obtain no hold whatever on the human mind.[29]

We use the phrase "collective action dilemma" to refer to situations in which the production of some group benefit is limited or prevented by the temptation to free ride. We use the phrase "collective action" to refer to cooperation aimed at overcoming the collective action dilemma. The collective action dilemma is also sometimes referred to as the collective action problem and the social dilemma.[30] The group benefits that result from collective action are referred to as public, collective, social, and common goods and as common-pool resources. These terms refer to slightly different things, but what they all have in common is nonexcludability, that is, it is difficult or impossible to prevent people from benefiting from them even if they have not participated in their production or paid some other cost in order to gain access to them. Public goods have the additional feature of being nonrivalrous, that is, one can consume them without reducing anyone else's ability to do so. Common-pool resources do not have that characteristic: those who consume them reduce others' ability to do so. Because collective action dilemmas often prevent public goods from being supplied, they are also sometimes referred to as public goods problems. Although public goods are often provided when people find ways to overcome the collective action dilemma, they may also be provided through other means. For example, an individual may choose to provide a public good despite any free riders who might benefit because of other benefits he or she receives from doing so, such as notoriety, acclaim, or a paycheck.[31]

Coordination and coordination problems

Collective action dilemmas stymie cooperation because there is a conflict of interest among those who stand to benefit from cooperating. Coordination problems also stymie cooperation, but for a different reason: lack of common information.[32] We use the phrase "coordination problem" to refer to situations in which people would benefit from cooperating but they lack the necessary information with which to do so. As with "cooperation," one way to understand this is to break the term down: "coordination." People "ordinate" similarly, that is, they prefer the same outcome, but they lack the information necessary to ensure it. Coordination is the type of cooperation that happens when this problem is solved.

Some scholars exclude coordination from the category of cooperation.[33] Given our broad definition of cooperation and the corresponding breadth of material we wish to cover in this book, we see no reason to do so. Indeed, the importance we see in coordination as an aspect of human cooperation is reflected in the title we chose for this book. As those familiar with social science research on cooperation will recognize, our title refers to Thomas Schelling's seminal work on focal points and coordination problems.[34] Schelling pointed out that solving coordination problems is made easier when people's attention is drawn to prominent or salient focal points. Schelling's classic example is the fact that most people in New Haven, Connecticut, will answer "Grand Central" when asked where they would go to meet someone else in New York City if no specific meeting place has already been arranged. As Schelling himself noted, this may reflect the fact that he asked people in New Haven. Grand Central Terminal is salient for people from New Haven because it is where their commuter trains arrive, but it may not be equally salient for everyone. Had Schelling asked people from New Jersey or Long Island, many might have answered "Penn Station," because that is where their commuter trains arrive in New York City. People's prior experiences frame the way they solve coordination problems and help them work together as teams. Progress in scientific research is also often a matter of coordination, of shared information. Our title reflects our hope that this book may act as a kind of focal point solution for scholarship on cooperation, providing some common ground upon which researchers from different disciplines and intellectual traditions can share ideas and build new theory.

Our approach

We approach these issues from the perspective of evolutionary theory (Lee) and political science and related social sciences (Beth). We believe that both perspectives have important contributions to make to the study of cooperation. With a few notable exceptions, most social scientists to date have not incorporated findings from evolutionary biology into their theories of human behavior. Evolutionary theory can provide a theoretically grounded set of expectations about what motivates humans to act. However, just as cooperativeness is not an explanation of cooperation, human motivations by themselves cannot provide a full explanation of human cooperation. Human decisions are affected by the environment in which those decisions are made. To use the language of game theory, the environmental context affects the payoff structure of those decisions. For modern humans, that environmental context is largely man-made. Man-made institutions—from cultural norms like kinship structures and food sharing customs to formal organizations like corporations and governments—may lead to cooperation even among individuals who are otherwise not at all predisposed to cooperate. Given how much cooperation among humans relies upon such institutions, our species's remarkable levels of cooperation may have as much (or perhaps more) to do with our ability to create institutions that work with and around our evolved propensities as they do with a history of natural selection for cooperativeness. Of course, our ability to create such institutions in the first place is due in large part to those same evolved propensities, so these two explanations—evolution and institutions—work together, as we will attempt to show in the chapters that follow.

Plan of the book

Chapter 2 is a primer on the concept of adaptation. Readers already familiar with evolutionary theory may choose to skim it, but those whose memories of the theory are rusty will find that it provides the background needed to understand later chapters. Following George Williams, we emphasize his idea that adaptation is a "special and onerous concept that should be used only where it is really necessary."[35] Adaptations benefit organisms, but so do many other things, and we need to be careful not to see adaptations in cases where such benefits are only fortuitous, not designed by natural selection. Our ears are, clearly, adaptations, designed to let us hear sound. Just as clearly, the fact that they also provide a convenient place to attach ornaments is merely a fortuitous by-product of selection in favor of ears, not an adaptation for earrings. The study of adaptations has an important but limited role to play in the study of behavior, including those involved in cooperation. Explanations involving adaptation focus on the forces of selection. Such forces do not shape behaviors in an immediate, mechanistic, or short-term way, but rather in a causally distal way, by favoring some genes over others, usually over quite long periods of time. Explanations of this type—which evolutionary biologists call "ultimate" or "distal"—complement explanations that deal with more immediate—or, in the jargon of evolutionary theory, "proximate"—causes. Proximate causes are typically the concern of the social and behavioral sciences. Understanding this division of labor among the sciences is an important step toward the creation of a unified—but not uniform—approach to behavior in general and cooperation in particular.

In chapter 3, we review the social science literature on cooperation, focusing primarily on the rich bodies of theoretical and empirical research on collective action that grew out of Mancur Olson's challenge. Despite Olson's skepticism regarding people's ability to solve the collective action dilemma, even he knew that people sometimes did so, and he identified some reasons why. Small groups have many advantages over large ones, such as the fact that in small groups each member is more likely to have enough of a stake in the outcome to make it worthwhile to contribute to the public good. Even large groups can make a go of it despite the free rider problem. For example, they can create a special set of benefits that only those who contribute to the public good are able to receive. Other scholars have emphasized the nonmaterial benefits that people derive from contributing to public goods. In some cases, entrepreneurial individuals take on more than an equal share of the costs of providing a public good in exchange for other kinds of benefits to themselves, such as money and prestige.

Chapter 4 examines the ways in which our evolved psychology helps us to cooperate. As models based on the Prisoner's Dilemma game have made clear, one key to cooperation is simply the ability to avoid cheaters, free riders, and other uncooperative types. Mounting experimental evidence shows that we are indeed quite good at distinguishing between those who have or are likely to cooperate and those who are not and at identifying people who break social rules. Given that others are on the lookout for cooperators, it follows logically that we also should be good at displaying our own cooperative tendencies. Again, a growing body of evidence shows that even a hint of an audience can make us more cooperative and that we care deeply about our reputations.

Chapter 5 focuses on the role that organizations play in making cooperation possible. Often, successful organizations are ones that build on the successes of previous organizations, using their abilities to mobilize and motivate people and to frame issues to leapfrog over the collective action dilemma. One of the most important areas of research to grow out of this tradition is the study of how groups of people successfully manage common-pool resources. A common-pool resource is one that is difficult to prevent people from using and that can be used up. Examples include fisheries, forests, and water. In economic jargon, such resources have low excludability and high subtractability or rivalry. At one time, it was commonly believed that the problem of managing a common-pool resource could be solved in only two ways: turn it into private property by dividing it up or use the power of the government to create and enforce a scheme for sharing it. Thanks to the work of people like Nobel laureate Elinor Ostrom, we now have many well-documented examples of common-pool resource management that take a middle road, one that maintains the resource as common property while specifying rules and procedures by which it can be managed and its productivity maintained rather than squandered. The existence of organizations and other kinds of groups sets the stage for a process known as "cultural group selection." Cultural group selection is distinct from and leads to different predictions about human behavior than biological group selection. Cultural group selection may help explain our flexible coalitional psychology.

Chapter 6 focuses on coordination problems, which we think should have a more prominent role in the study of cooperation. Coordination problems are essentially problems of information: Although people would benefit from coordinating their activities, they lack common knowledge about how to do so. Even worse, they may actually have common knowledge about how to solve the problem but not know it. In Michael Chwe's words, they lack "common metaknowledge."[36] Thomas Schelling recognized one way to overcome this problem that we have attempted to capture with the title of our book: focus on prominent, salient focal points that others are also likely to focus on. If you know you are to meet someone in New York City, but you don't know where to go and have no way to reach the other person, why not try Grand Central Terminal at noon? While Grand Central may not be the most convenient place for either you or the person you are meeting, it is prominent enough that there is a good chance that it will occur to both of you. The value of coordinated social action may have been great enough among our ancestors to have led to the evolution of special abilities, such as the ability to imagine what is in someone else's mind, that make it easier for us to solve coordination problems.

Coordination is also the theme of chapter 7, which focuses on how social interactions can lead to the spontaneous emergence of norms, conventions, and other social institutions that help coordinate social behavior. Because an interaction with one individual is, in a sense, an encounter with everyone that he or she has previously encountered, a network of interacting individuals can pool information and lead to the rapid generation of a consensus without a central plan of any kind. The study of emergent social phenomena has a long history in the social sciences. It was particularly important during the Scottish Enlightenment of the eighteenth century, with Adam Smith's "invisible hand" metaphor being the most widely known (and most widely misunderstood) manifestation of the idea. Modern social and behavioral scientists are using the logic of emergence to phenomena ranging from improvisational comedy to riots to traffic jams.

Chapter 8 summarizes our findings and offers some observations regarding the relationship between the social and life sciences. In our view, although different sciences explain different phenomena, they share at least one common goal: consilience. In other words, their explanations should, at the end of the day, mesh with one another. This creates a network rather than a hierarchy among the sciences and their knowledge claims. Consilience is made possible by emergence. The emergent properties of living things necessitate the existence of a discipline—biology—and body of theory designed to explain them that is dependent upon and consilient with but distinct from physics and chemistry. Similarly, the emergent properties of social phenomena necessitate the existence of social science disciplines that are dependent upon and consilient with the natural sciences while remaining distinct from them. The study of cooperation provides an excellent example of how this division of labor among the sciences can lead to a wide range of complementary insights regarding specific social phenomena.

The first step on this journey is to ask a deceptively simple question: What kind of organism are we dealing with, anyway? Of course, we all think we already know the answer. After all, we are all humans, and the fact that we all have a good intuitive understanding of what makes people tick is demonstrated every day by our successful interactions with one another. The key contribution that evolutionary biology can make to the study of cooperation is to provide an understanding of who we are as organisms that is rooted in something more than mere intuition and experience. Because the starting point for any application of evolutionary biology to behavior is the concept of adaptation, it is the focus of the next chapter.

Box 1.1. Experimental Economic Games

Experimental economic games are an important source of data for the study of cooperation. Many such games have been devised in order to examine different aspects of cooperation. Here we describe the games that have been used most commonly in both laboratory and field studies. The names of the games and other terminology associated with the games are for the benefit of researchers. Research participants are not usually told the name of the game they are playing. In most experiments, players are anonymous to each other. The reason for anonymity is simple: without it, researchers would have a hard time determining which aspects of the players or their relationships with one another are responsible for the way they play the games.

The Dictator Game involves two players. The first player is given some amount of money and can allocate none, some or all of it to the second player. Sometimes the "second player" is a charity of some kind rather than an individual. The Dictator Game is often used to assess generosity. Technically, the Dictator Game is not a game because only one player's payoff depends on what the other player chooses to do. It is usually referred to as a game simply because it is used alongside other games.[1]

The Ultimatum Game also involves two players.[2] The first player (often called the proposer) is given some amount of money and can allocate none, some, or all of it to the second player. The second player (often called the responder) can accept or refuse the offer. If the offer is accepted, then both get to keep whatever they have at that point. If the offer is rejected, neither player gets anything. A common variation, called the "strategy method," is to ask the second players what their response would be to a range of possible offers from the first player rather than using only the first player's actual offer. Researchers often use the Ultimatum Game to assess participants' sense of fairness and their willingness to punish at some cost to themselves. It is common for researchers using the Ultimatum Game to refer to offers of 50 percent as "fair" and offers of greater than 50 percent as "hyperfair." However, there is nothing inherently "fair" about 50 percent. Such terms imply both that the researchers know what their subjects consider "fair" and that their subjects chose to give a particular amount either because they thought it was fair or because it would be perceived as fair by the responder. Because researchers rarely have that level of knowledge about their subjects' motivations, the terms

[1] Forsythe et al. 1994.
[2] Güth et al. 1982.

Box 1.1 continues

Box 1.1 continued

"fair" and "hyperfair" should be eschewed in favor of "half" and "more than half." Because the first player has an incentive to avoid rejection by the second player, low rates of rejection in the Ultimatum Game may reflect a consensus among players in a particular setting for what constitutes a "fair" offer. Conversely, high rates of rejection might reflect a lack of a consensus among players in a particular setting regarding what constitutes "fairness" in the Ultimatum Game. High rejection rates can also arise for other reasons. Players among the Au and Gnau of highland Papua New Guinea, for example, appear to bring a familiar cultural frame to bear when they play the Ultimatum Game. Like many peoples in highland New Guinea, the Au and the Gnau have competitive gift-giving systems in which a large gift creates an obligation for an even larger one to repay it at some time in the future. Accordingly, Au and Gnau players tend to both make and reject large offers in the Ultimatum Game.[3]

The Trust Game is another two-player game. Usually, both players are given equal initial endowments so that fairness is not foremost in their minds. A variation is to give only the first player an initial endowment. The first player can give any proportion, including none or all, of his or her endowment to the second player. The experimenter then multiplies that amount by some number, usually two or three. The second player can then give any portion of his or her money back to the first player. Both players would end up with the most money if the first one gives everything to the second one, who then returns half to the first player. The game gets its name from the fact that such a strategy requires the first player to trust the second one. The Trust Game is also sometimes referred to as the Investment Game, with the first player referred to as the "investor" and the second player as the "trustee." It is often used to assess participants' senses of trust, obligation, and gratitude.[4]

The Third Party Punishment Game involves three players. The first two players play a Dictator Game. The third player, having observed the behavior of the first player in the Dictator Game, is given money with which he or she can punish the first player. Typically, the amount that the third player chooses to spend on punishment is doubled and then that amount is subtracted from the amount taken home by the first player. The punisher gets to keep whatever he or she does not spend on punishment. For obvious reasons, this is often used to

[3] Tracer 2003.
[4] Berg et al. 1995

Box 1.1 continues

Box 1.1 continued

study participants' willingness to punish and their motivations for doing so. It is also possible to add punishment to other two-person games.[5]

The Public Goods Game involves multiple players, typically three to five. All players are given equal initial endowments and can then put any portion of them into a common pot. The experimenter then multiplies the amount in the common pot by some number, often two, and then divides it equally among the players. Players get to keep whatever they kept from their original endowments plus their share of the common pot. A variation is to add a punishment round. In the punishment round, anonymity is maintained but players' contributions are revealed. Players can then use some of their money to reduce the amounts received by others. The usual procedure is for the experimenter to double anything spent to punish and then reduce the punished player's take-home amount accordingly. The Public Goods Game is used to study collective action dilemmas.[6]

[5] See, for example, Fehr and Fischbacher (2004).
[6] Ledyard 1995.

CHAPTER 2

Adaptation

A SPECIAL AND ONEROUS CONCEPT

IMAGINE THAT YOU ARE A FLYING FISH. Thanks to your long pectoral fins, you can escape from predators by skimming the tops of waves. Clearly, those long fins of yours are an adaptation, designed by natural selection to provide a particular benefit. But returning to the water is also beneficial. Although the water is home to the predators from which you fly, it is also where you will find good things such as food and mates. Indeed, returning to the water is so beneficial that, were you a well-educated fish, you might be tempted to apply the concept of adaptation not only to your ability to fly but also to your eventual return to water. But you would be wrong. Evolutionary theory and the concept of adaptation are unnecessary here. Plain old gravity will do the job.

The flying fish example comes from George Williams's *Adaptation and Natural Selection*. Williams's goal was to refine the concept of adaptation and to clarify when it is appropriate and, perhaps even more importantly, when it is not. Adaptation, he argued, "is a special and onerous concept that should be used only where it is really necessary."[1] Adaptations are features of organisms that have been designed by natural selection to provide specific kinds of benefits. We can recognize an adaptation by the ways in which its design reflects the purpose for which it was selected.

Our goal in this chapter is to explain the concept of adaptation and how it is applied (and sometimes misapplied) to cooperation. Because cooperation is so common in our species, it must often be more beneficial than noncooperative alternatives. Is it, therefore, an adaptation? No, not in the strict sense of the term. The term "cooperation" refers to a very wide range of behaviors that occur in a wide range of circumstances for a variety of reasons, making it inappropriate for the concept of adaptation. Although cooperation is not itself an adaptation, it is often the outcome of other, more specific adaptations. Some adaptations that support cooperation were favored by selection precisely because of their ability to do so. For example, our ability to empathize improves our ability to cooperate by making it easier for us to recognize when others share our goals, and this benefit is likely to have been part of its selective advantage.

In other instances, the cooperation that we find in contemporary societies is built upon adaptations that were originally favored by selection for other reasons. The market, for example, is an enormous and long-lasting system of cooperation that arises from the self-interest of individuals rather than from adaptations favored by selection because of their contributions to cooperation. Either way, identifying the adaptations that help us work together is the first step toward understanding evolutionary theory's contribution to the study of human cooperation.

Adaptation and levels of explanation

Although the study of adaptations is an important part of biology, it is certainly not the discipline's only job. Other biologists study the nitty-gritty details of how organisms work, how they develop, and what their ancestors were like. This division of labor was spelled out decades ago by evolutionary biologist Ernst Mayr and animal behaviorist Niko Tinbergen and is now a basic feature of how students are educated in the evolutionary sciences.[2] These levels of explanation help clarify the relationship between evolutionary and nonevolutionary approaches to society, culture, and behavior, creating an opportunity for a mutually beneficial division of labor between them in their efforts to explain cooperation and other phenomena.

For any phenomenon, there are four complementary levels at which it can be explained. First come proximate explanations. By proximate, we mean "causally close to the phenomenon at hand." Proximate explanations are good answers to "how" questions such as "how does the eye work?" A proximate explanation of the eye would focus (pardon the pun) on how it is supposed to work—cornea, lens, retina, and so on. Because such explanations are typically rather mechanistic, the things they explain are often referred to as "proximate mechanisms." A little farther away from the phenomenon in terms of causal distance lies the developmental or ontogenetic level of explanation. If we examine how the eye develops in the zygote, fetus, child, and so on, then we are offering a developmental explanation, one that is complementary to the mechanistic proximate level explanation. The next level answers questions about why things work the way they do, and not some other way, by focusing on how the forces of natural selection shape adaptations. For example, what selection pressures caused the eye to be designed as it is? This is often called the "ultimate" level of explanation. Unfortunately, that label has a tendency to sound like a claim of significance and thus a challenge to other explanations. Because "ultimate" explanations are no more significant or important than explanations at other levels, we and many other evolutionary scientists prefer to call them "distal." Distal also contrasts nicely and logically with "proximate" in a way that "ultimate" does not. Finally, if we focus on the raw materials that selection acted upon in order to design the eye, which includes everything from the earliest light-sensitive cells to the hair follicles that produce eyelashes, we are offering a phylogenetic explanation. The important thing to remember is that while there may be competing hypotheses at all these levels, correct explanations at the different levels are complementary to one another.

Social and behavioral scientists concentrate mostly on proximate explanations. This is understandable not only because proximate explanations often have great practical usefulness but also because, in our everyday lives, the causes and effects of which we are most immediately and keenly aware are those that occur at the proximate level. However, when reading the social science literature on cooperation, evolutionary scientists are often frustrated by an apparent lack of interest in the other levels of explanation. Many psychological explanations of behaviors, for example, begin and end with the identification of what motivated them. An evolutionary scientist sees this as an important part of the picture, but not the whole picture. Where did that motivation come from? Was it favored by natural selection? What were the raw materials that natural selection worked with to shape that particular motivation? The same can be said of explanations emerging from economics and political science in terms of "utility." Such explanations are fine at the proximate level, but they cannot be the whole story. If humans find "utility" in something, why is that so? Did natural selection play a role?

Perhaps the greatest frustration emerges from explanations that purport to be at one level but really belong at another. This has an unfortunate tendency to occur at the interface between evolutionary and social scientific work on cooperation. Evolutionary theorist Stuart West and his colleagues have identified several examples in the scholarly literature on cooperation in which proximate explanations are passed off as distal ones.[3] One particularly clear-cut case of confusion about levels of explanation appears in an article by Dominique de Quervain and his colleagues on the neural basis of altruistic punishment. First, they ask a "why" question, one that should lead to an explanation at the distal level: "Why do people punish violators of widely approved norms although they reap no offsetting material benefits themselves?"[4] Then, they offer an explanation at the proximate level: "We hypothesize that individuals derive satisfaction from the punishment of norm violators." Well, perhaps they do. But even if that is true, it does not serve as an explanation of how a tendency to find satisfaction in the punishment of norm violators could have evolved.

Adaptations, fortuitous benefits, and by-product mutualism

Williams's flying fish example demonstrates an important constraint on the concept of adaptation: the fact that an organism experiences a benefit does not necessarily mean that an adaptation is involved. Like flying fish returning to the water, the organism may simply be taking advantage of an aspect of nature that is not a product of its own evolution. In short, the benefits that adaptations are designed to provide must be distinguished from whatever benefits an organism receives fortuitously. In addition, benefits provided by adaptations must be distinguished from whatever benefits an organism provides fortuitously to others. To illustrate this second point, Williams used the example of apples.[5] Apples, he wrote, provide humans with a wide variety of benefits, ranging from "Newtonian inspiration" to improving the economy of Kalamazoo County. But, clearly, apples exist not due to the benefits they fortuitously provide humans but rather due to the benefits they provide the trees that produce them: reproduction and seed dispersal. Because apple design is a result of selection in favor of characteristics that improve their ability to provide benefits to trees, apples can rightfully be identified as an adaptation of the apple tree.

Apples and other fruits provide food to frugivores not because they are altruistic, but rather because they need for their seeds to be dispersed. Producing fruit is costly to the plant, but worth the reproductive benefit it provides. The fact that fruiting also benefits frugivores is simply a by-product. Similarly, frugivores eat fruit not in order to disperse seeds but rather to obtain nutrients. From their point of view, seed dispersal is simply a side effect. Biologists call this "by-product mutualism."[6] Because this particular example involves two parties that exchange mutualistic benefits as by-products of their behaviors, we can call also it "pseudoreciprocity."[7] When by-product mutualism involves one party providing a resource, such as a large prey item, that it shares simply because it cannot afford to defend it from others, the result may be referred to as "tolerated theft" or "tolerated scrounging."[8] Because by-product mutualism does not involve organisms working together, it does not fit the definition of cooperation we gave in the previous chapter. Nevertheless, by-product mutualism is an important aspect of the study of cooperation because it may help explain many cases of public goods provisioning and apparent cooperation.[9]

By-product mutualism is an idea that has been reinvented multiple times in both the social and life sciences. Economists, for example, capture the same idea by referring to benefits that are provided incidentally as "positive externalities."[10] Many of the theories of collective action described in the next chapter involve aspects of by-product mutualism. The Snowdrift Game provides a simple model of such a situation, in which one individual might choose to provide a benefit to another simply as a side effect of his or her own self-interested actions.[11] The scenario is simple: Two cars are stuck in a snowdrift. Somehow the snow has been arranged so that one driver cannot dig out his own car without also enabling the other driver's car to go free. If both drivers help dig the cars out, they both get the benefit of being able to drive their cars. They also both pay the cost of digging, but it is only half of what it would cost one of them to do the job alone. If one digs while the other sits in his car, they both get the benefit but one incurs the entire cost of digging. The upshot is that digging produces a benefit for oneself, but it also provides one for someone else, who has the option of simply accepting the benefit or doing something to earn it. Unless there is some other way in which the two parties are connected (e.g., as kin or members of the same community), then whether natural selection would favor digging depends solely on the cost-benefit ratio to the individual, not on the benefit that might be incidentally provided to someone else.

Adaptation and selection: Natural, artificial, social, and sexual

What does it mean to say that something was "designed" by natural selection? One thing that it definitely does not mean is that there is an actual "designer," intelligent or otherwise. Selection designs organisms, but not through any sort of conscious design or plan. Charles Darwin's great breakthrough was his understanding that different members of the same species, though similar in important ways, also vary. Some of those variations have an impact on how many offspring their bearers ultimately have—their *reproductive success*—and can also be inherited by their descendants. If some variants are more beneficial than others—i.e., if their bearers leave more descendants behind—then a species will change over time. The retention of variants that enhance reproduction and the loss of those that do not lead to the development of characteristics, some simple and some complex, that help organisms deal with the challenges they face in their struggles to survive and reproduce. Such characteristics are adaptations.

Darwin called this process natural selection. He included the adjective "natural" in order to emphasize both the similarities and differences between it and artificial selection, which is what people do when they breed plants and animals. Those apples in Kalamazoo County, for example, are the results of not only millions of years of plant evolution but also centuries of deliberate breeding. Darwin pointed out that a breeder does consciously what nature does automatically: reward some varieties with greater reproductive rates than others. Other frugivores had been inadvertently doing the same thing—choosing to eat some apples (and apple precursors) over others and thus helping to spread their seeds—for millions of years before humans came along and started to selectively breed them. While a particular design is a conscious and deliberate goal in artificial selection, natural selection blindly influences the ways organisms fit their environments, doing so with no goals and no foresight.

Because genes replicate, they—or, to be more precise, the information they contain—are potentially immortal. Organisms, in contrast, are comparatively short-lived and ephemeral. Therefore, what drives evolution and the design of adaptations is differential reproduction, not differential survival. However, because organisms must survive for at least a short time in order to reproduce, the purpose of many adaptations is to enhance survival. They may do so in a variety of ways—improving the organism's ability to obtain nutrients, improving its ability to avoid predators, improving its ability to resist pathogens, improving its ability to work with others to accomplish common goals, and so on.

Many adaptations are responses to selection pressures imposed by a species' physical environment (e.g., the climate) and members of other species (e.g., predators, prey, and pathogens). But members of one's own species can also act as a selective force. In its broadest formulation, this is referred to as "social selection."[12] Paradoxically, social selection may drive the evolution of characteristics that may seem rather antisocial. For example, cats hiss and arch their backs when threatened. Such threat displays evolved as ways to display a willingness and an ability to defend oneself and one's territory. However, social selection may also have been an important force behind adaptations that support cooperation. For example, humans are remarkably good at empathizing with others, at imagining what others are thinking, and at identifying others who might be good cooperative partners, all of which help us achieve our high levels of cooperation.[13] Virtually all models of the circumstances that might have favored adaptations that support cooperation implicitly include social selection, though few scholars have focused on it explicitly.

When social selection has an impact on an organism's ability to acquire mates, it is referred to as sexual selection. As Darwin first pointed out, sexual selection can be a powerful force, producing some of nature's most remarkable adaptations—peacocks' tails, rams' horns, and so on. Humans and other species in which both parents often help care for offspring have an incentive to choose mates not only on the basis of their physical characteristics but also on the basis of their personality and behavior. Are they willing and able to care for offspring? More to the point, are they willing and able to do so in a way that complements one's own parenting abilities? Thus, sexual selection may have contributed to the evolution of the ability to coordinate one's behavior with someone else's and to focus on a shared goal. In this way, sexual selection, in combination with biparental care, may have contributed to the evolution of cooperativeness in our species.[14]

Phylogeny: Adaptation's raw materials

Adaptations are built upon the raw materials provided to natural selection by an organism's phylogeny or evolutionary history. As a result, an organism's evolution is path dependent: what it is like now is a function of what it was like in the past, and what it is like now constrains what it can become in the future. A full evolutionary account of an organism's adaptations includes an understanding of not only how it has been modified by natural selection but also its phylogeny. Thus, as we mentioned in chapter 1, evolutionary biology is both a well-developed theoretical science and a historical science. Although some evolutionary biologists emphasize one over the other, both are necessary for a full understanding of how organisms have been designed by the forces of selection.

An easy way to appreciate how phylogeny constrains natural selection is by spotting flaws in how organisms are designed. Although organisms are remarkably well designed to deal with the problems they routinely face, such flaws are not hard to find. Consider the human foot. In many ways, it is remarkably well adapted for our form of bipedal locomotion, carrying one's weight smoothly from heel to big toe with each step. In other ways, it is deeply flawed. Soft and full of nerves, it compares poorly with an antelope's hoof or an elephant's foot as an appendage for moving across the ground quickly and without injury. As Williams facetiously asked, "Why is man a mere biped and not a centaur?"[15]

The answer lies in the similarities between the human foot and the human hand. Like the hand, the foot has digits, soft surfaces, and many, many nerves. Our foot's resemblance to a hand reflects the fact that until recently it was even more like a hand than it is now. This was undoubtedly helpful to our arboreal ancestors, who used their feet as well as their hands to grasp branches. The human foot is an excellent compromise between the raw materials available to selection—a very hand-like foot—and the need for steady and rapid locomotion across the land. But it is still a compromise, as is every adaptation.

The foot is an evolved example of what computer scientists and other engineers call a "kludge." Although the term's etymology is obscure, its meaning is made clear if we treat it as a misspelled acronym for "clumsy, lame, ugly, dumb, but good enough." Selection creates kludges. Although they may work remarkably well, they are still kludges. Even the eye is a kludge. Although our eyes work remarkably well, they are also poorly designed. When the cells in the retina are stimulated by light, the signals they send to the brain must travel within the eyeball and out a hole in the back of the eye. The result is a reduction in the amount of light that reaches the retina and a blind spot where the optic nerve is located. To neuroscientist David Linden, even the human brain can best be seen as a collection of kludges, and those kludges make us what we are: "It's not that we have fundamentally human thoughts and feelings *despite* the kludgy design of the brain as molded by the twists and turns of evolutionary history. Rather, we have them precisely *because* of that history."[16]

Although behaviors do not fossilize, we can learn about their phylogenetic histories by comparing across species. Although comparisons are frequently made between humans and our closest ape relatives, some adaptive problems are so ancient and widespread that comparisons between humans and even very distantly related species can be enlightening. For example, our ancestors have had the problem of selecting high-quality mates for as long as there has been sexual reproduction. It is not surprising, then, that we find similarities in mate preferences across a very wide range of species. A preference for symmetry in mates, for instance, is found among humans as well as among species as distantly related as the zebra finch.[17] The reason may be that symmetry indicates high-quality genes: If an individual has what it takes to develop symmetrically despite all of the problems presented by its environment, then its genes must do a good job of coding for the characteristics that allow it to deal with that environment. Behavioral characteristics may also be important in mate choice, particularly when the relationship lasts beyond fertilization and both members of the pair contribute to offspring care. Humans are one of the few primates that form long-lasting pair bonds and that have biparental care of offspring. Although the role that biparental care played in the evolution of human pair bonding is still debated,[18] it is reasonable to suppose that our ancestors sought mates who not only were physically attractive but also showed signs of being cooperative, committed, and caring. Thus, some adaptive problems related to cooperation may be very ancient indeed, and the forms that cooperation takes among humans may reflect this deep evolutionary history.

Selection routinely finds new uses for old adaptations. The ape foot, so well designed for life in the trees, was put to a new use in our ancestral line: terrestrial locomotion. Though birds' wings were clearly designed for flight, they have found new uses among such flightless birds as penguins (swimming) and ostriches (display). Evolutionary biologists Stephen Jay Gould and Elisabeth Vrba argued that such repurposed adaptations should be called "exaptations."[19] However, virtually every current adaptation is constructed out of past adaptations.[20] Before birds had wings and apes had feet, their tetrapod ancestors had forelimbs and hindlimbs. And before that came fishes with fins, and so on.[21] When we do want to emphasize the way in which an adaptation has been repurposed, we will refer to it as a "co-opted adaptation."[22]

As with feet and wings, the repurposing of old adaptations can be the result of natural selection over evolutionary time. The co-optation of old adaptations can also happen much more quickly as human institutions emerge and adjust, whether through deliberate planning or through their own evolutionary processes, to make use of our species' evolved psychology. The repurposing of adaptations through sociocultural processes, rather than through natural selection, may play an important role in the high levels of cooperation seen among contemporary humans. For example, we evolved in a world dominated by kinship relationships, and although we no longer live in that world, we often make creative use of kinship, even among nonkin, in order to encourage cooperation. The reason why humans and other organisms should be expected to favor kin over nonkin was spelled out by the great evolutionary biologist William D. Hamilton. He set out to solve a basic evolutionary puzzle: If selection acts through the differential survival and reproduction of individuals, why do individuals sometimes do things that harm their own chances of surviving and reproducing while enhancing those of others? Hamilton's answer, which is usually referred to as kin selection or inclusive fitness theory, focused on the fact that individual organisms share genes with genetic relatives. A tendency to behave altruistically toward others will spread in a population if it is coupled with a sensitivity to relatedness. A tendency to help close relatives will be favored over both individual selfishness and a tendency to help distant relatives and nonrelatives. The result will be exactly what we see among humans and other social animals: mechanisms for distinguishing kin from nonkin, social behaviors that differ according to degrees of genetic relatedness, and social worlds structured by kinship.

Because humans, unlike other animals, routinely cooperate with nonrelatives, it is often said that relatedness is of only limited use in explaining human cooperation. But it should not be dismissed too quickly. If our goal is to explain why living humans are inclined to cooperate with nonrelatives, it is worth remembering that they are the products of an evolutionary past in which nonrelatives were encountered much less often than they are in contemporary, urban societies. Even today, in small-scale hunter-gatherer, pastoralist, and horticultural societies, day-to-day interactions occur mainly among kin, and so, too, does most cooperation.

This evolutionary history makes it possible to create what Peter J. Richerson and Robert Boyd have called a "work around," that is, the use of adaptations that arose in other circumstances to solve new problems.[23] Making people more cooperative by triggering their kin recognition mechanisms is surprisingly easy to do. Using photos of faces that had been manipulated to resemble either her experimental subjects or someone else, Lisa DeBruine showed that people are more trusting of people with faces similar to their own.[24] Similarly, people are more likely to help strangers who happen to share their last name.[25] This can also be accomplished through the use of kinship terms. Kin terms in political rhetoric (e.g., "brother," "sister," "motherland," "fatherland," and so on) both are common and increase persuasiveness, and religious organizations that demand celibacy often use fictive kin categories to structure their relationships.[26] Organizations that train suicide bombers use kin terms to manipulate and motivate their recruits.[27] Kin recognition mechanisms and associated emotions may be triggered in more subtle ways by the bonding effect of shared struggle, trauma, and triumph, as among initiates in secret societies (e.g., fraternities and sororities), athletic teammates, and soldiers.[28]

Adaptation and environments, old and new

Adaptations are designed to solve specific problems found in specific environments. The environment in which an adaptation was designed is its environment of evolutionary adaptedness, often abbreviated as EEA.[29] A common error is to apply the EEA concept to all of a species' adaptations collectively rather than to specific adaptations. For example, some early work in human evolutionary psychology was premised on the idea that our species' EEA was roughly equivalent to the Pleistocene.[30] To reduce the likelihood of such an error, some now refer to the adaptively relevant environment, or ARE, rather than the EEA.[31] "Adaptively relevant environment" emphasizes the idea that each adaptation originated in a particular environment. The ARE (or EEA) for the human foot, for example, was the African forests and savannahs of a few million years ago, when our line split from that of the chimpanzees and bonobos and started to become more bipedal, to about half a million years ago, by which time the foot had developed its current form.[32] Other aspects of our current phenotype are associated with more ancient environments. For example, a characteristic we share with all other amniotes (i.e., mammals, reptiles, and birds) is the fact that our offspring develop inside amniotic sacs and other protective membranes. Those membranes solve the very ancient problem of reproducing outside water, where our amphibian and fish ancestors developed. Yet other aspects of our current phenotypes may have originated in relatively recent environments. For example, humans whose ancestors ate starchier diets have more genes for an enzyme (amylase) that helps break down starches than those of us whose ancestors ate relatively low-starch diets. Because the amount of starch in our diets increased rapidly when we domesticated plants, the ARE for this trait is likely to be quite recent in evolutionary terms.[33]

Like feet, amniotic sacs, and other features of our physiology, aspects of human psychology and behavior may have been designed by natural selection to solve adaptive problems associated with environments in our evolutionary past. Some aspects of human mate preferences, such as the preference for symmetry mentioned earlier in this chapter, are likely to be very deep-seated because they deal with a problem that is so ancient. Some aspects of the psychology of parenting (e.g., the formation of emotional bonds between mothers and offspring) and sociality (e.g., a general preference for kin over nonkin) are also likely to be quite ancient and therefore quite deep-seated and phylogenetically widespread. Other aspects of our psychology, including some that may play a role in cooperation, may have evolved in more recent environments. For example, human psychology may be particularly well suited to the formation and maintenance of coalitions. The fact that chimpanzees also form coalitions suggests that this ability may go back at least a few million years,[34] but its accentuation and elaboration in humans may reflect the importance of cooperation within coalitions among our ancestors.[35]

Organisms are not merely passive recipients of the environments in which they live: they also actively shape them, often in ways that benefit themselves and their offspring. This process has been labeled "niche construction." Even very simple organisms engage in niche construction. The earthworm, for example, shapes and improves its environment (the soil) by eating and excreting it. Many other animals make their environments a bit more to their liking by creating structures—nests, dens, burrows, and so on—in which to live and raise offspring. Such modifications of the environment can have effects that persist over generations, creating a sort of ecological inheritance that runs in parallel to and interacts with genetic inheritance.[36] Humans are champion niche constructors, not only modifying their physical environments in radical ways but also constructing social environments that powerfully shape human life.[37] One way to view the social sciences is as the study of human niche construction. As we will show in later chapters, social scientists study the ways in which humans shape their environments and how they in turn act within and react to those environments.

Because adaptations are associated with specific past environments, there is no guarantee that they will work well when environments change.

This is known as the problem of novel environments. Because humans have been changing animals' environments very rapidly, examples of the novel environment problem are easy to find. Anyone with a large window has surely heard the occasional "thunk!" when a bird slams into it. Or consider the wild animal carcasses that litter the shoulder of a typical highway. Animals evolved to avoid predators, not vehicles coming at them at speeds beyond what even a cheetah can reach, and so they often react to cars and highways in ways that would have been adaptive for their ancestors facing a predator but that simply don't work in the new environment. The armadillo is a particularly good example of this phenomenon. Its natural defense is rather odd. When confronted by a predator, an armadillo will leap straight up into the air. This may startle a coyote, but it does not deter an oncoming semi. The result: dead armadillos.

In many ways, humans now live in evolutionarily novel environments, as well. Whereas our ancestors lived as hunter-gatherers in small-scale societies, surrounded by relatives, we live in large-scale societies, dependent upon agriculture, industry, trade, and a complex division of labor, routinely interacting not only with nonrelatives but often with complete strangers. Interestingly, many of the changes that led to our novel environment hinged upon our ancestors' abilities to find new and increasingly effective ways of cooperating with one another. The state and the market stand out as evolutionarily unprecedented forms of cooperation, but even subsistence farmers must coordinate their efforts in ways that go beyond the relatively small-scale cooperation seen among hunter-gatherers.[38]

Just as highways present armadillos with an evolutionarily unprecedented problem, highly artificial environments may trigger responses in humans that would have been adaptive for our ancestors but that are not adaptive for humans in the situations in which they now sometimes find themselves. For example, sweet, salty, and fatty foods, when eaten in moderation, provide important nutrients, so selection favored those among our ancestors who found them tasty and sought them out. Given that in our ancestral environments such foods were rare, overindulgence was not much of a risk. In today's environment, of course, it is easy to overindulge in such foods, and doing so is both tempting and bad for one's health. Similarly, adaptations that were favored by selection because they supported adaptive cooperation among our ancestors may not necessarily produce adaptive behaviors in all circumstances in which humans now find themselves. Highly artificial environments may trigger responses that would have been adaptive for our ancestors but that are no longer so. Laboratory behavioral experiments, for example, correspond with nothing experienced by our ancestors. Given that such experiments are evolutionarily unprecedented, we should not be surprised that they often produce results (e.g., generosity that cannot be reciprocated) that seem to run counter to our evolutionary theoretical expectations and counter to the best interest of the subjects of the experiment. Like armadillos leaping in front of cars, such behaviors may be easy to trigger because they would have made good sense in the kinds of small-scale, kinship-based societies in which our ancestors lived.

Adaptation, judgment, and misjudgment

An inappropriate or even maladaptive response to a situation may also represent not a flaw in an adaptation but actually one of its design features. This is the idea behind error management theory, sometimes referred to as the "smoke detector principle."[39] An ideal smoke detector is one that goes off only if a dangerous fire is spreading. Given that it is impossible to design such a perfect smoke detector, which would you rather have in your home: a smoke detector that will definitely go off if a dangerous fire is spreading but that might also go off if you, say, burn a piece of toast, or one that will definitely not go off if you burn your toast but that might fail to go off if a dangerous fire is spreading? The key is that errors may come in many forms, but their costs may not all be equal. If some kinds of errors are much more costly than other kinds, then it might be worth it to make a few of the cheaper ones in order to avoid the really costly ones.

Evolutionary psychologists such as Martie Haselton, David Buss, and Randy Nesse have argued that this same design feature may be built into many adaptations. Thus, many male organisms are easily aroused sexually because the reproductive cost of nonarousal is so great and the cost of a misfire so low. Human males may be quick to interpret female friendliness as an indication of sexual interest, even when it is not meant that way, because the cost to our ancestors of missing a mating opportunity would have been so much greater than the cost of being turned down. People may be prone to mistake sticks for snakes, but not vice versa, because the cost of ignoring a snake is so much greater than the cost of ignoring a stick. The same logic has been applied to belief in supernatural beings and superstitions: a tendency to make causal inferences that are incorrect but low cost (e.g., ghost detection when the bushes shake) may be favored by selection if it also leads to the ability to make correct causal inferences that are highly beneficial (e.g., predator detection when the bushes shake).[40]

Political scientist Dominic Johnson has extended this reasoning to help explain human cooperation. He reasons that a tendency to behave cooperatively due to a fear of supernatural retribution may be adaptive, despite the cost of behaving cooperatively when one could get away with free riding, if it leads people to avoid costly punishment for free riding at the hands not of the gods but of their fellow humans.[41] The logic of error management theory may also apply to decisions we make about cooperation in a more general way. If extreme suspiciousness or selfishness would lead an organism to forego opportunities to benefit greatly from cooperation, then selection might favor a tendency to trust others despite the occasional cost of being exploited by the untrustworthy.

Adaptation, culture, and language

In everyday life, "culture" means opera, the cinema, and eating paté with *cornichons*—in other words, what is sometimes called "high culture." Anthropologists, in contrast, use culture as a technical term, devoid of the judgmental baggage associated with such notions as "high culture" and "low culture." Unfortunately, that is just about the only aspect of culture on which anthropologists can agree. E. B. Tylor got the ball rolling in the nineteenth century by defining culture as "that complex whole which includes knowledge, belief, art, morals, law, custom, and any other capabilities and habits acquired by man as a member of society."[42] Since Tylor's time, anthropologists have proposed dozens of different definitions of culture. Most (though by no means all!) have some merit, but many are also confusing or misleading in various ways.

One of the most common flaws in anthropologists' definitions of culture is a failure to distinguish culture from behavior and its products, both material (e.g., tools) and nonmaterial (e.g., social institutions). Such definitions lead to circularity and confusion when culture is then used to explain behavior. If all behavior is considered culture, then it becomes impossible to distinguish between behaviors that are shaped by culture and behaviors that aren't, and, ironically, culture loses its explanatory power. By claiming to be an explanation of all behavior, culture so broadly defined ends up unable to explain anything at all. As a remedy, we define culture as *socially transmitted information*.[43] This is an ideational definition, that is, one that separates culture from behavior and its products, both material ones (e.g., tools) and nonmaterial ones (e.g., social institutions). Thus, the information one needs to bake a cake (the recipe, knowledge about techniques and ingredients, etc.) is culture, but the act of baking a cake is not, and neither is the cake itself.[44] Although ideational definitions of culture originated in cognitive and symbolic anthropology,[45] they are now the norm in evolutionary anthropology and animal behavior studies, and they are essential to models of culture trait evolution.[46] Ideational definitions of culture enable us to avoid circular explanations of behavior and generate genuine explanations.[47]

Although it is common in the social sciences for culture to be invoked as an explanation of behavior without much—or any—consideration of alternatives, that is not the case in the evolutionary sciences. Among animal behaviorists, evolutionary anthropologists, and evolutionary psychologists, culture is every bit the "special and onerous concept, that should be used only where it is really necessary" that adaptation was to Williams. Indeed, animal behaviorists were for many years very reluctant to attribute any nonhuman behaviors to culture, but most are now willing to accept it as a necessity when other factors, such as genetic and environmental differences, cannot explain a pattern of behavior. Social scientists would benefit from a similarly skeptical and demanding attitude toward culture as an explanation of behavior, treating it as a hypothesis to be tested rather than as an a priori assumption.

Culture is sometimes seen as a novel element of the human environment. It is certainly true that humans have a great deal more culture than does any other species. But is culture a novel element of our environment? Not entirely. Animal behaviorists have shown that culture, at least in simple forms, can be found in a very wide range of species. Guppies, for example, mimic one another in a variety of ways, including mate preferences and routes to food sources.[48] Rats learn from each other how best to process pine cones as food.[49] Birds and whales learn to sing songs, which vary from place to place.[50] Innovations in feeding methods spread among whales in ways that cannot be explained genetically.[51] Chimpanzees and other primates have a large number of socially transmitted traditions regarding such things as tool manufacture, tool use, food acquisition, and social behavior.[52] The list of such examples is growing rapidly, and greater sophistication in the detection of animal culture is likely to lead to the discovery of many more.[53] It is now clear that we need to take into account socially transmitted information—culture—not only when studying humans but also when we try to explain many nonhuman behaviors.

Given that culture is so widespread, it is also likely to be quite ancient. In that sense, it is no more a novel element of our environment than are plants or predators. We already know that culture has been around long enough for it and genes to have coevolved.[54] Some examples of gene-culture coevolution among humans have been well documented. Adult lactose absorption—the ability to drink fresh milk without ill effects—has evolved within the past few thousand years in at least two human populations with cultural traditions of dairying, one in Northern Europe and the other in Africa. Lactose, a sugar found in milk, provided those populations with an additional source of calories and, in northern latitudes, a substitute for vitamin D.[55] Most adults elsewhere in the world cannot digest lactose, though some can consume cheese and yogurt because much of the lactose they contain has been predigested by bacteria. Another example of gene-culture coevolution comes from sickle-cell anemia, a genetic disorder that is favored in some environments because in its heterozygous form (i.e., when an individual has one gene for it but not two) it confers some resistance to malaria. When malaria-carrying mosquitoes multiply for what is ultimately a cultural reason—the clearing of tropical forest for farming—the sickle-cell trait also becomes more frequent.[56]

Lactose absorption and sickle-cell anemia show how culture can lead to changes in the physical environment that can then lead to changes in gene frequencies. But culture itself is also an element of the environment. Just as the physical environment (pathogens, predators, climate, etc.) can lead to changes in gene frequencies, so can culture create selection pressures that reward those genes that best equip their bearers to deal with culture. In short, culture may favor a variety of psychological adaptations. Because this has been going on for millions of years, humans and nonhumans are likely to share many culture-related psychological characteristics. One possible example may be the simple tendency to mimic those who are in a position to have more accurate or more up-to-date information than oneself. This is seen not only in guppies that mimic one another's mate preferences and route choices,[57] but also in humans, who do very similar things.[58] Such similarities may be homologies (i.e., characteristics present in multiple species due to shared ancestry) or analogies (i.e., characteristics present in multiple species due to independent evolution). Given how ancient and widespread simple forms of culture seem to be, we find it both more plausible and more parsimonious that they are homologies rather than analogies. However, a complete test of this idea has yet to be conducted.

Although culture as a broad phenomenon is not a novel element of our environment, many aspects of culture are found only among humans. Indeed, thanks to our ability to use symbols and language, we are able to spread culture in ways and amounts not seen in any other species. Many specific culture traits are also novel in the sense that nothing like them is seen in any other species. The list of uniquely human culture traits is long, ranging from the trivial (e.g., we are the only species that smokes tobacco) to the profound (e.g., we are the only species that controls fire). This reflects the fact that one of culture's most adaptive qualities is its ability to generate novelty, which enables us to adjust to a wide range of circumstances without undergoing genetic change. Thus, while culture does have the potential to lead to circumstances that are so novel that we are poorly equipped to deal with them, its flexibility is more commonly put to use in adaptive ways. This is true of not only the innovative technologies we use to deal with the physical environment but also our creativity in dealing with our social environments. Culture allows us to fashion new social arrangements, including ones that promote cooperation, using the raw materials provided by evolution.

A specific cultural innovation, one that is unique to humans, is language. Language's importance to our ability to cooperate with each other cannot be overestimated.[59] In addition to being enormously helpful to the coordination of social behavior, language is the key to the role that reputation plays in encouraging people to be cooperative and provides people with a means to signal their commitments to group endeavors and values. All of this is well known and widely recognized. Indeed, probably because social scientists rarely compare humans with nonhumans, language and its many benefits for cooperation are taken for granted in most of the social science literature on the topic. Less well recognized is the role that language plays in helping us think in ways that go beyond the prelinguistic and nonlinguistic cognitive mechanisms that predate language and that we probably share with many non-humans. For example, the ability to make rough comparisons of quantities has been demonstrated in some nonhumans as well as among humans who speak languages that lack words for specific quantities.[60] But languages that do provide words for exact quantities give their speakers "a new route for the efficient encoding of experience"[61] that sets the stage not only for the entirety of modern mathematics but also for mathematical concepts, such as market prices, that are crucial to the successful functioning of institutions that help humans cooperate. Language has a similar ability to enhance our ability to think and communicate clearly about a variety of other realms of experience, including color, navigation, and the perception of emotion,[62] all of which can improve the chances that cooperation will arise by making it easier to share ideas and coordinate social behavior.

Genes, individuals, and groups: Adaptations at multiple levels?

Although we have emphasized George Williams's careful delineation of the concept of adaptation, most biologists remember *Adaptation and Natural Selection* more for its critique of the concept of biological group selection. Biological group selection is the idea that selection acts most strongly through the differential survival and reproduction of entire groups of organisms. Biological group selection is important for us to address in this book because if it were common, then explaining altruism, generosity, cooperativeness, and other prosocial traits would be much easier than it is when we limit ourselves to selection at lower levels, such as individual organisms and genes. If it were common, and if cooperative groups were more successful than noncooperative groups, then we should expect cooperation to be unproblematic and ubiquitous. If biological group selection is either uncommon or not powerful, then we will need a host of other clever ideas to explain why competition among individuals or genes can lead to individuals and genes that sometimes do things that hinder their own efforts to reproduce while helping others to do so, that is, why they sometimes act in ways that appear to be altruistic.

One reason to give biological group selection serious consideration is the undeniable fact that it has played an important role in some of the major transitions in the evolution of life on this planet. For example, selection at the genetic level favors individual genes that replicate at the highest possible rate relative to other genes, and there are indeed many interesting examples of truly "selfish genes."[63] However, if there were only selfish genes, then we would still live in some sort of primordial soup of replicating DNA or RNA molecules. Or, rather, *we* would not live at all. Organisms like us are able to exist because selection pressure at the level of the genome has led to the evolution of mechanisms that make life hard for genetic free riders. For example, with sexual reproduction comes the potential for a scramble among genes to make it into gametes, those all-important lifeboats to the future, because each one contains only half an organism's nuclear DNA. Meiosis, the process of cell division that leads to gamete production, creates an even playing field among the genes in a genome by giving each one an equal chance of making it into each gamete. By pooling the fates of all of the genes in a genome, meiosis produces a genetic version of John Rawls's "veil of ignorance," a situation in which all members of a group have an equal chance of receiving a benefit that, by its very nature, not all can receive.[64] Although some rogue "driving" genes do manage to cheat their way into more than half of an organism's gametes, the vast majority of genes follow the rules, making it into an average of half an organism's gametes. If biological group selection can produce an adaptation as successful as meiosis, it is only natural to wonder whether it could also be responsible for other important adaptations.

When Williams wrote his book, thought about selection at different levels was often either fuzzy or wholeheartedly focused on group- and even species-level benefits of adaptations. For example, Alfred E. Emerson, a prominent zoologist at the time, suggested that dying from old age evolved due to its benefits to the species, such as a shortening of generation times that would make it easier for a species to evolve in response to changing conditions.[65] Another prominent biologist, Alexander Carr-Saunders, marshaled evidence to show that "primitive" human societies use abortion, infanticide, and sexual abstinence to avoid overpopulation, arguing that because such behaviors are costly to individuals, they must have evolved due to their group-level benefits.[66] Williams found these scenarios unlikely, and he tried to set evolutionary theory straight.

The clearest statement of the biological group selectionist position was offered by V. C. Wynne-Edwards, who argued that many common animal behaviors, such as territoriality, flocking, and the formation of dominance hierarchies, could be explained as adaptations at the level of the group, not the individual.[67] In this view, species form dominance hierarchies in order to limit the number of individuals who get to reproduce, thus keeping a check on population growth. To Williams, in contrast, a dominance hierarchy is simply "the statistical consequence of a compromise made by each individual in its competition for food, mates, and other resources. Each compromise is adaptive, but not the statistical summation."[68] We may marvel at the many-leveled pecking orders present in many nonhuman societies, but they are merely the result of each individual attending to his or her own personal social situation, not selection in favor of the hierarchy as a whole.

In addition to declaring adaptation to be "a special and onerous concept," Williams argued that "it should be attributed to no higher a level of organization than is demanded by the evidence."[69] And Williams did not see much evidence that would demand biological group selection. On the contrary, Williams argued that adaptation, whether physiological or behavioral, human or nonhuman, can be explained by "the simplest form of natural selection, that of alternative alleles in Mendelian populations."[70] In other words, adaptations arise not because of competition between different groups but rather because of competition between different genes. Ten years later, biologist Richard Dawkins captured the essence of this idea with the metaphor of the "selfish gene," and selfish gene models have been the basis of the evolutionary study of behavior ever since.[71]

Williams did not deny that biological group selection could occur, at least in principle. He even included the only instance of biological group selection in nature that was well documented at the time he wrote, one involving some unusual genes in mice.[72] Since that time, some other examples of biological group selection in nature have also been identified (e.g., sex ratios in cooperatively breeding spiders).[73] But the scarcity of such examples illustrates Williams's point: selection at the level of individual organisms is almost always a stronger force than selection at the level of groups, making biological group selection highly unlikely and even "impotent" compared to the strength of selection at lower levels.[74]

To understand Williams's argument, imagine selection working on a trait in opposite directions at the levels of individuals and groups. For example, selection at the level of individuals pushes them toward ever greater reproduction. However, that could conceivably threaten the group's survival by leading to overpopulation, crowding, and food shortages. If overpopulation is bad enough, it could lead to a population crash and the extinction of the entire group. But if some group were to develop a way of limiting individuals' reproduction, it would survive longer than groups that fail to come up with such a group-level adaptation. For example, individuals in such a group might forego reproduction unless they controlled a large territory—larger, that is, than they would need to simply live and reproduce at the high rate favored by selection at the level of the individual. That would limit the number of breeders in the population and so limit population growth. If groups that developed such mechanisms for controlling population growth survived longer than groups that failed to develop them, then this would be an example of selection at the level of the group. Thus, if selection at the level of the group is strong enough, it could indeed shape group-level adaptations despite the pressure of selection at the individual level to keep reproducing at a high rate.

As plausible as this may sound, it does not occur very often in nature. Why not? A couple of years before Williams wrote his book-length critique of biological group selection, evolutionary biologist John Maynard Smith identified the key problems in Wynne-Edwards's biological group selectionist logic.[75] In order for biological group selection to work, extinction of entire groups needs to be a real threat in much the same way that death for an individual organism is a real threat. In most circumstances, however, it is not. Population explosions rarely lead to starvation and group extinction because other things, such as increased predation, usually limit population growth before such a crash can occur. Another reason has to do with migration rates. In order for this kind of biological group selection to occur, groups need to be distinct from one another. Migration rates between them need to be low. Otherwise, individuals who don't go along with the group-level adaptation can make it from group to group and out-reproduce the individuals who display the group-level adaptation. Because the extinction of entire groups is rare and because there is usually some migration between groups, biological group selection is rarely a very powerful force.

That is where the argument stood when Williams's book was published in 1966, and, for many evolutionary biologists, anthropologists, and psychologists, that is where it still stands. Some have even gone so far as to declare biological group selection a "fallacy,"[76] but, as we have seen, to do so is itself a fallacy. The question is not whether biological group selection is possible. It is. The question is whether it matters. In the words of Maynard Smith, "Group selection will have evolutionary consequences; the only question is how important these consequences have been. If there are genes which, although decreasing individual fitness, make it less likely that a group (deme or species) will go extinct, then group extinction will influence evolution. It does not follow that the influence is important enough to play the role suggested for it by some biologists."[77]

Among many students of social behavior, however, biological group selection is still taken quite seriously. There are many reasons for this. One source of biological group selection's continued popularity in some circles is a failure to distinguish between a trait's evolved function—why it was favored by selection—and its fortuitous ancillary benefits. Consider, for example, this quote from an article proposing a biological group selectionist explanation of egalitarian sharing rules found in some Papua New Guinea societies: "Such sharing norms are beneficial for the *group* because they insure group members against the uncertainties in *individual* food acquisition success."[78] If this kind of reasoning were valid, then anything that benefits individuals would be a product of biological group selection because it would also benefit the groups to which they belong. Just as it is easy to mistake the fortuitous side effect of gravity on the behavior of flying fish for an adaptation, it is easy to mistake benefits experienced by groups via their constituent individuals for adaptations at the group level. Although groups may indeed experience benefits because of the behaviors of their constituent individuals, the most likely evolutionary explanations for those behaviors rest on individual-level selection, not biological group selection.

Another, better reason for a continuing interest among scholars in biological group selection is an important article published in 1970 by George Price.[79] In a single equation, Price captured selection's ability to act simultaneously at different levels. The Price equation has a structure reminiscent of a statistical method known as analysis of variance, or ANOVA. Whereas ANOVA partitions variance in a dependent variable into within-group and between-group components, the Price equation partitions the effects of selection on changes in the frequency of a gene into within-group and between-group components.[80] By including selection at all levels within the same framework, the Price equation puts biological group selection alongside individual-level selection as a possible, if rare, source of evolutionary change. The theoretical possibility of selection at multiple levels is now often referred to as "multilevel selection."[81] Although it would be fair to say that we are all multilevel selectionists now, a close reading of *Adaptation and Natural Selection* shows that we always were. For most evolutionary biologists, all that has really changed is the terminology.

Yet another reason for biological group selection's resurgence is a combination of theoretical work and energetic salesmanship by its advocates, particularly biologist David Sloan Wilson. For more than three decades, Wilson has waged a Stakhanovite struggle on behalf of the biological group selectionist cause. Many of his arguments were captured in an influential book, coauthored with philosopher Elliot Sober, titled *Unto Others*.[82] Readers of *Unto Others* were told that the tide in evolutionary biology was shifting toward biological group selectionism, and that even such eminent theorists as William D. Hamilton were embracing the idea. Hamilton was the author of multiple breakthroughs in evolutionary theory, but none were more important than the idea that altruism toward kin could evolve through a process of gene-level selection. Hamilton's goal was to explain altruism without resorting to biological group selection. His demonstration that biological group selection was not required for altruism to evolve was a major blow to the theory. According to Sober and Wilson, however, Hamilton had a change of heart regarding biological group selection after learning of the Price equation, an event recorded in a "neglected" publication.[83] There, Hamilton showed that the Price equation could be used to recast his inclusive fitness model in terms of multilevel selection. But did Hamilton really change his mind regarding the importance of biological group selection? No. In fact, his main argument in that same publication was essentially the same as Williams's: "I shall argue that lower levels of selection are inherently more powerful than higher levels, and that careful thought and factual checks are always needed before lower levels are neglected."[84] Was this article "neglected" by scholars? According to the Web of Science, it has been cited more than three hundred times. While that is far fewer citations than many of Hamilton's other publications have received, most scholars would love to have their publications be so neglected.

Sober and Wilson's argument rested in large part on an unusual definition of "group." Previously, the definition of "group" had corresponded, more or less, with common sense. To Williams, it was clear that "group" referred to "something other than a family and to be composed of individuals that need not be closely related."[85] To Maynard Smith, a "group" had to be reproductively isolated from other groups, as would be the case with a deme (i.e., a population of organisms) or a species.[86] For Sober and Wilson, in contrast, a "group" exists any time two or more organisms interact, however briefly, in a way that has an impact on their fitness. Thus, any two individuals who interact in a wide variety of ways—mating, fighting, competing, cooperating, and so on—and then go their separate ways constitute a "group." This "ultra-liberal definition of group"[87] allows Sober and Wilson to argue that almost any social interaction is subject to biological group selection. This muddying of the terminological waters is not helpful. As Maynard Smith has argued, the words we use do matter: "A group selection terminology leads us to look for factors causing a difference between variation within and between groups; a kin selection model leads us to look for relatedness; and a game theory model leads us to look for frequency dependence and non-additive fitness inter-actions."[88] If we were to accept Sober and Wilson's definition of "group," then nothing would be left out, and the very idea of "biological group selection" would lose whatever explanatory power it might otherwise have had. By purporting to explain everything about social life, it would explain nothing.

Nevertheless, examples like meiosis that show biological group selec-tion's ability to shape important adaptations—even if only occasion-ally—make it hard for many scholars to give up on the idea that biologi-cal group selection helps explain human cooperation. Perhaps meiosis itself is a starting point. If we could find human groups with similar mechanisms, we would have good support for the theory that our minds and our social behaviors have been shaped by the forces of biological group selection. Do any human societies have institutions that resemble meiosis? David Sloan Wilson and Elliot Sober think so.[89] The Hutterites are an Anabaptist Christian community well known for their communal-ism and high fertility, with colonies scattered across the North American High Plains. Due to their high fertility, Hutterite colonies tend to grow, and eventually they must divide, with some colony members moving else-where to establish a new colony. As Wilson and Sober explained, colony members are chosen for the group staying and the group leaving in a way that will result in groups that are about the same in terms of size, age, sex, skills, and "personal compatibility. The entire colony packs its belongings and one of the lists is drawn by lottery on the day of the split." Wilson and Sober concluded that "the similarity to the genetic rules of meiosis could hardly be more complete."

Actually, the analogy between meiosis and Hutterite colony fissioning customs is weak. Meiosis does not sort genes according to how well they get along with each other, and the well-documented association between relatedness and "personal compatibility" among humans makes it un-likely that any Hutterite community is a random sample of the commu-nity from which its members originally came. In fact, research conducted long before Wilson and Sober hit upon the Hutterite example has shown quite the opposite. In 1964, geneticist Arthur P. Mange examined related-ness among Hutterites before and after colony divisions. The result: fis-sioning increases relatedness. This is reflected in people's surnames. In many of the colonies Mange studied, only a few surnames were repre-sented. In one colony, everyone shared a single surname.[90]

How common is a community fissioning rule that even roughly re-sembles meiosis? Although Wilson and Sober suggested that "the Hut-terite social organization is not unique but represents a fairly common type of social organization in ancestral environments," we know of no evidence that any other human group uses a fissioning procedure that resembles meiosis even as remotely as the Hutterite custom. Instead, communities tend to fission along lines of kinship, friendship, and reciprocity. Napoleon Chagnon's detailed work on village fissioning among the Yąnomamö of Venezuela, for example, demonstrated the important role that relatedness plays in decisions about where to move and with whom.[91] Chagnon's student James Hurd documented the same phenomenon among Old Order Amish, who, like the Hutterites, are Anabaptists.[92] Although the ostensible causes of fission in Amish congregations are usually matters of doctrine, Hurd documented that people sort themselves into new congregations mostly on the basis of relatedness. The same fissioning pattern has been observed not only among humans, but also among nonhuman primates.[93] As Wilson and Sober acknowledged, if Hutterite social organization is unique then "it could not be interpreted as an evolved adaptation." And indeed it is not one. The Hutterite community fissioning rule is, rather, an example both of how ingenious people are and, thanks to its uniqueness, of how hard it is to get people to act like Hutterites.[94] If we really were the products of biological group selection, Hutterite-like examples would be commonplace, cooperation would be more common than it is, the collective action dilemma would not be such a dilemma, and we probably would not be so motivated to write this book.

More recent advocates of the idea that human psychology and behavior have been shaped by the forces of biological group selection point not to the Hutterites but rather to findings from a large number of studies using experimental economic games of various kinds (see Box 1.1). In Dictator Games, people often give to the other party even though they could maximize the amount they take home by giving nothing. In Ultimatum Games, low offers and even high offers are sometimes rejected despite the fact that this means that neither player will go home with anything from the game. In Public Goods Games, people often contribute rather than free ride. In Third-Party Punishment Games, people often use some of their own resources to punish others. These and other examples of altruism have been given the label "strong reciprocity." "Strong reciprocators" are "predisposed to cooperate with others and punish non-cooperators, even when this behavior cannot be justified in terms of self-interest, extended kinship, or reciprocal altruism."[95] Because the costs of "strong reciprocity" cannot be recovered by the individual strong reciprocator, the implication is that it must have evolved through biological group selection.

"Strong reciprocity" has been criticized on several different grounds, including but not limited to the idea that it is a product of biological group selection.[96] However, perhaps the most basic problem with "strong reciprocity" is the term itself. "Strong reciprocity" is neither reciprocity nor "strong." By definition, the costs of "strong reciprocity" cannot be recovered through reciprocity or any other means. It makes no more sense to call unreciprocated generosity "strong reciprocity" than it does to call it "generalized reciprocity," as anthropologist Marshall Sahlins once proposed.[97] If it is not reciprocity, then it should not be called reciprocity (see Box 4.1). The reason for calling it "strong" is also unclear. Perhaps its "strength" lies in its ability to be maintained despite its costs to the individual actor.

Setting aside the shortcomings of "strong reciprocity" as a technical term, the question is whether we need biological group selection to explain the behaviors it refers to, that is, altruistic giving and altruistic punishing. Because so much of the support for "strong reciprocity" comes from experimental economic games, many critics have focused on those games and how their results are interpreted. Most such games are played in laboratories, often with college students, though they have also been used by many anthropologists in field settings around the world. Regardless of where the games are played, players are typically anonymous to each other and the game, even if it is run for multiple rounds, is essentially a one-shot experience. Given how little such games resemble real life, it is not surprising that players respond to them in inappropriate ways. Altruistic giving and altruistic punishment appear to be hothouse flowers, quickly wilting when experimental conditions change even slightly.[98] For example, while people do often give away money in the Dictator Game, it is possible to nearly eliminate such donations by making people earn the money they have to give.[99] This effect is enhanced if complete anonymity is maintained, with the identities of generous and stingy people unknown even to the experimenters.[100] Furthermore, experimenters can turn generous dictators into selfish hoarders simply by giving them the additional option of taking funds away from the other player.[101] Similarly, donations in Public Goods Games and rejections of offers in Ultimatum Games may be artifacts of the game-playing method itself. It takes time for people to learn how games work, and, until they have a chance to learn the game thoroughly, they are likely to make use of whatever framing cues they notice. In Public Goods Games, for instance, donations typically start high and diminish over subsequent rounds as it becomes clear that others are free riding. Because Ultimatum Games have no single Nash equilibrium (a situation in which neither player can benefit by changing his or her strategy), they may be particularly difficult to learn.[102]

Even people who have had plenty of time to learn how a game works can make mistakes. In an ingenious study designed to make it easy to detect errors, Rolf Kümmerli and his colleagues ran subjects through Public Goods Games, with a twist: the common pot was multiplied not by a factor of two, as is usually done, but rather by a factor greater than the number of players in each game.[103] This modification made it in everyone's best individual interests to contribute everything they had to the pot. Even when players had received extensive training in how to play the game, had played several rounds of the game, and were told that anything that they held onto would not remain with them at the end of the game, some did not contribute everything they had to the common pot, and many players viewed the other players more as competitors than collaborators. This has implications for "strong reciprocity" in that some of the behaviors observed in experimental games that have previously been used as support for it, such as nonzero contributions in Public Goods Games, may simply be errors.

While it is true that Third-Party Punishment Games often elicit punishment, punishment in such games is more common in large societies than in small ones.[104] This has at least two implications for the idea of "strong reciprocity." First, it suggests that punishment that promotes cooperation in small societies is mainly by second parties (i.e., the party originally offended), not by third parties (i.e., observers). Given that our ancestors lived in small societies, this casts doubt on the relevance of Third-Party Punishment Game results to the question of how humans evolved to be so cooperative. Second, it suggests that the key to third-party punishment rates may be the institutional structures in which game players live. Large societies typically include police and other deliberately created institutions whose task is to act as third-party punishers. In such societies, third-party punishment may be more likely to occur to players as a script for game playing because they see it in action every day. Punishment also diminishes when players have the option to retaliate against punishers.[105]

The most parsimonious explanation for the altruism observed in experimental games is that it is an example of the novel environment effect explained earlier in this chapter. Thus, as Terrence Burnham and Dominic Johnson have pointed out, the term "strong reciprocity" is no more necessary than "strong sperm bank cuckoldry" would be as a label for infertile couples' use of another man's sperm, together with such modern technologies as cryogenics and artificial insemination, to have children.[106] Humans did not evolve in a world of one-shot anonymous interactions and laboratory experiments. Our ancestors lived in small-scale societies in which there was precious little privacy and virtually always an audience,[107] and it is that "evolutionary audience"[108] that we should keep in mind when contemplating the behaviors of modern humans. Advocates of "strong reciprocity" argue that experimental subjects are surely aware of their anonymity,[109] but the awareness of people on a roller coaster of their own absolute safety does not prevent them from being scared out of their wits.[110] An additional "evolutionary audience" to keep in mind is the actors themselves. If economist Robert Frank is correct that the best way to signal one's commitment to cooperation is to be cooperative regardless of the audience present, then it should come as no surprise that some experimental subjects find it difficult or even impossible to behave completely selfishly.[111]

Williams recognized that biological group selection's primary appeal was not scientific, but rather aesthetic. It makes people feel good to think that organisms have been designed in such a way as to make them helpful and generous to others. " [P]opulations in which individuals, such as worker bees, often jeopardize their own well-being for a larger cause" are more appealing than "those whose members consistently act only in their own immediate interests," and "those in which individuals normally live in peace or active cooperation and mutual aid" are more appealing "than populations in which open conflict is more in evidence."[112] When applied to humans, such thinking becomes not only aesthetically but also politically appealing to many people. How much easier it would be to create a world in which people are generous and cooperative if we *had* evolved at the group level. This may explain why, unlike most evolutionary biological explanations of human behavior, theories of biological group selection have had an easy time finding admirers in the social sciences.

By now it should be clear that we are deeply skeptical about the relevance of biological group selection for the study of cooperation. However, it would be misleading for us to report that there is a consensus on this issue among scholars. There isn't. But perhaps we can all agree with Robert Kurzban and Athena Aktipis's call for an examination of "the footprints of multilevel selection."[113] Their argument is simple and appealing: Let's accept the fact that selection acts on multiple levels and move on to an examination of the evidence. Are there adaptations that are plausible products of biological group selection and that cannot be adequately explained in terms of selection at lower levels? Do adaptations have features that are designed to benefit groups and that don't do so simply fortuitously? Or can we explain what we see, whether in the study of human cooperation or elsewhere, in terms of selection acting at levels other than the group? While we share Williams's, Hamilton's, and Maynard Smith's skepticism regarding the strength of biological group selection in most circumstances, we also agree with Kurzban and Aktipis: where there is a testable hypothesis, we should examine the evidence.

Adaptation and justification

Social scientists avoid evolutionary explanations of human behavior for a variety of reasons. One common concern is that an evolutionary explanation of a behavior might somehow morally justify the behavior. For the study of cooperation, this might seem like a minor concern. After all, cooperation is a good thing, isn't it? But what about the cooperation that makes warfare possible, the cooperation that occurs in organized crime, or the cooperation among price-fixing firms in a market economy? If we argue that such behaviors have evolutionary explanations, are we justifying them in some moral sense? Of course not. To argue that something is morally good or acceptable merely because it is in some sense "natural" is known as the "naturalistic fallacy." That phrase was coined by philosopher G. E. Moore over a century ago,[114] but it was based on a much older and broader principle laid down by David Hume in the eighteenth century: one cannot derive a statement of how things should be—an "ought" statement—from a description of how things are—an "is" statement.

This fallacy is well known to evolutionary scientists, and they are always on the lookout for it. They understand that simply because something was designed by natural selection, it is not necessarily "good" for anything other than enhancing an organism's genetic representation in future generations. However, this principle seems to be less widely known in the social and behavioral sciences. In those fields, it is common to see scholars use explanations of why behaviors occur as moral justifications of those behaviors. Why do some people in the world modify (some would say mutilate) their daughters' genitalia? The answer "because their culture tells them to" is, to some, a moral justification of such behavior.[115] But the logical leap from "is" to "ought" is the same as that made by people who violate the naturalistic fallacy by claiming that a behavior occurs for a reason and therefore is justified morally. The same error is present whenever behaviors are thought to be morally justified because they have a particular cause, whether it is because culture leads people to do them, because they make good economic sense, or because they are politically effective. The fact that evolutionary explanations rely upon ideas like natural selection and adaptation do not make them any more vulnerable to this kind of error than explanations of behavior based on any other causal factor. Therefore, to avoid or dismiss an evolutionary explanation of a behavior on the grounds that it may be misinterpreted as a moral justification of that behavior makes no more sense than to avoid cultural, economic, psychological, political, or sociological explanations of that behavior.

From Williams to Olson

At about the same time in the mid-1960s that George Williams was writing *Adaptation and Natural Selection*, Mancur Olson was working on a parallel track in the social sciences. In writing *The Logic of Collective*

Action, Olson's goal was quite similar to Williams's: clear up an area of theoretical confusion regarding why people sometimes do and sometimes do not form groups to pursue their common interests. Like Williams, Olson eschewed models that assumed that group-level benefits explained individual-level behaviors and focused instead on individual-level benefits. In the next chapter, we explore Olson's arguments, those of his supporters, and those of his detractors.

CHAPTER 3

The Logic of *Logic*, and Beyond

EACH YEAR, 136 million people are born while fewer than half that number die. If this rate of growth continues, in just sixty-three years the world's population will double, while the world's resources remain finite. Overpopulation has been blamed, rightly or wrongly, for deforestation, air pollution, water pollution, global warming, urban sprawl, and the fall of ancient civilizations.[1] And yet, what can one person do to change this situation? Any single person declining to have children will have statistically zero impact on the overpopulation problem. If the person had hoped to enjoy a life full of children and grandchildren, then the personal cost of such a move would be enormous. The emotional costs of foregoing a traditional family are all borne by the individual who gives up a dream of having children, while the effect of this individual sacrifice on the actual world population is negligible. Assuming that fears of overpopulation are well founded, then this is a collective action dilemma in the extreme.

The mission of the nonprofit citizen group Population Connection is to attempt to address this problem. Boasting more than 110,000 "members, supporters, and participating educators,"[2] the Washington-based organization provides lesson plans to schools and lobbies Congress and the White House regarding overpopulation and unsustainable development. About ten years after the publication of Mancur Olson's treatise on why large groups such as this are so hard to form successfully, sociologists Harriett Tillock and Denton Morrison decided to use the group, then called Zero Population Growth, to examine the limitations they saw in Olson's theory.[3] They surveyed more than three hundred members of ZPG, asking them about the reasons they had joined the group, the size of their chapters, and how effective they believed the group and their own efforts to be. Not only did members of the group not receive any personal gain from joining the group, they also reported that they felt no social pressure to join. Instead, the benefit that members ranked most highly was the hope of advancing the collective good of achieving population control. Tillock and Morrison concluded that their results showed that Olson had been wrong. The tendency to free ride was not exhibited by the members of ZPG and that group had indeed successfully formed.

The editors of the journal gave Olson the chance to respond to Tillock and Morrison's article. Olson took issue with the researchers' methodology, especially the way that they had selected subjects to interview. The problem, he said, was that they had interviewed only members of ZPG and not any of the millions of people in the world who supported the organization's goals but did *not* join to help the organization achieve those goals. He wrote, "Your sample excludes all the members of the group who make non-deviant choices."[4]

Is the glass half empty or half full of irrational, "deviant" people? Is cooperation really as difficult as Olson suggested? Are there ways to overcome the collective action dilemma? To answer these questions, let us first look at Olson's *The Logic of Collective Action* in a bit more detail. Then we will give an overview of the extensions to his argument and the many ways of subverting his logic that have been documented in the social science literature.

A closer look at Olson's *Logic*

Olson's argument begins with the observation that, by definition (see Box 3.1), the members of a group have interests in common.[5] After that, Olson's logic and that of his predecessors diverge. Traditionally, analyses of groups had supposed that, because people in groups have things in common, groups must naturally form whenever there are interests in common. Before Olson, most scholars had assumed that if there were a need, a group would arise to address it. For some scholars, this idea was based on a vague and rarely specified supposition that there is "a fundamental human propensity to form and join associations."[6] Others saw the human propensity to form groups as an outcome of a process of societal and cultural evolution, with "states, churches, the larger business firms, universities and professional societies" performing many functions that are handled by kinship in more primitive societies.[7] For political scientist David Truman, it was simply a fact of social life that groups would arise to satisfy social needs: "With an increase in specialization and with the continual frustration of established expectations consequent upon rapid changes in the related techniques, the proliferation of associations is inescapable."[8] Olson pointed out that all of these views implicitly assumed "that private groups and associations operate according to rules entirely different from those that govern the relationships among firms in the marketplace or between taxpayers and the state."[9] Olson argued that such an assumption is wrong: "Unless the number of individuals in a group is quite small, or unless there is coercion or some other special device to make individuals act in their common interest, *rational, self-*

BOX 3.1. TYPES OF GROUPS

Mancur Olson's definitions of different types of groups are keys to understanding the social scientific literature on collective action. Here we present some of his more important definitions, along with a few from other sources.

Category: People who share a characteristic, such as red hair, blue eyes, or a fondness for chocolate. Although categories are not groups, people who belong to the same category do sometimes use their common characteristic as the basis for forming a group. Labor unions, which are formed by individuals who happen to be employed in the same industry, are a good example.

Group: Olson defined a group as "a number of individuals with a common interest."[1] Although this blurs the distinction between a group and a category, it was useful for Olson because it allowed him to go on define other types of groups, such as latent, large, and small groups, without worrying about whether they had actually moved beyond the status of category and developed some kind of internal structure. A somewhat more standard (and stringent) definition of group in the social sciences would be a number of individuals "who recurringly interact in an interconnected set of roles."[2]

Interest group: An organization or institution that makes policy appeals to government.[3] Because all interests that are not universally shared are "special," interest group scholars avoid the terms "special interest" and "special interest groups." Some interest group scholars prefer the phrase "organized interests" to "interest groups."

Large and small groups: Olson defined "large" and "small" groups not in terms of their actual size, but rather in terms of their potential size. "All people who breathe air" is a very large group (or, in non-Olsonian terminology, category) because all members of it could benefit from improved air quality. Such large groups may share important interests (e.g., reducing air pollution), but their size makes it unlikely that they will ever organize. Large groups are thus, in Olson's view, likely to be *latent groups* rather than actual ones. "All people who breathe air in a single office area" is a small group and, as such, is much more likely than a large group to organize to accomplish its common goals (e.g., elimination of secondhand cigarette smoke). For this reason, small groups are usually also, in Olson's terminology, *privileged groups.*

[1] Olson 1965:8.
[2] Keesing 1975:10.
[3] Baumgartner and Leech 1998:xxii.

*interested individuals will not act to achieve their common or group in-
terests*."[10] Beginning with the basic assumption that no sensible person or
organization would willingly expend more energy to create a group benefit than the person or organization received in return from that group
benefit, Olson predicted not only that groups would not form for the
common good but that in fact groups formed to advance the interests of
individuals, not those of groups.

Olson's logic was an economic logic, based on the behavior of firms in
the marketplace in their quest for profits. He noted that in the marketplace, every firm has an incentive to set prices as high as possible, in order
to make more money. If every company that made chocolate bars agreed
to set the price of chocolate at $10 a bar, all of the companies would
benefit from that high price. Any consumer who wanted chocolate would
have no choice but to pay the high price because there would be no other
source of chocolate (of course, the consumer could simply stop eating
chocolate, but if having chocolate was important enough, the high price
would have to be paid). The problem is that each company would have
an incentive to renege on the high price agreement because any company
that lowered its price would immediately capture most of the market. If
one company lowered its price, then others would follow, and soon the
collective benefit of high prices would be no more.

Olson extended this logic of the market to human social behavior. The
logic aims to explain why things that seem to be in the collective interest—called public goods—are often hard to achieve. Public goods, by
definition, represent benefits that, if attained, are shared not only by those
who worked to attain them but also by those who did not. A manufacturer, for example, benefits from a tariff on competing foreign goods regardless of whether that manufacturer contributed to the lobbying effort
for the tariff. All people benefit from breathing cleaner air regardless of
whether they joined an environmental group. Because nonparticipants
cannot be excluded from such benefits, any collective action faces a problem of free riders. It is easier (or in economic terms, less costly) to refuse
to participate, let others provide the good, and then partake of the benefit. Of course, if everyone did this, many collective goods would not be
provided, and Olson in fact predicted that "the provision of the collective
good will be strikingly suboptimal and the distribution of the burden will
be highly arbitrary."[11] This is exactly what happened in the case of Zero
Population Growth, whose members are a very small percentage of the
world's population and who have an infinitesimal direct effect on the size
of that population. With more than six billion people in the world,
110,000 people choosing to each have two or fewer children is barely a
drop in the bucket.

Olson's own solutions: Coercion, small groups, selective benefits, and by-product theory

Olson laid out several possible solutions to the problem of free riding and the possibility that no group would ever form. All of the solutions hinge on the assumption that no rational actor (whether it is an individual or an organization) would work to achieve a collective good unless its *own* share of the benefits from the collective good is worth more than the effort that individual or organization expended to achieve the collective good. That is, the benefits to the participant must always exceed the costs to the participant. Otherwise there is no incentive for the potential participant to join in. If the benefits don't exceed the costs, the participant will change into a nonparticipant—a free rider—and will not help to provide the collective good. The would-be participant will choose not to join the environmental movement, the union, or the boycott. To solve this dilemma, either the benefit to the individual of joining the collective action or the cost of not joining must be increased.

As Olson pointed out, coercion—force or the threat of force—is one very common way in which the cost of not joining is increased. But the use of force can be costly not only for those on the receiving end but also for those who are applying the force. After all, people have a tendency to fight back. Why, then, should one person stick his neck out by being the one who coerces others into joining the collective action? Why not free ride on others' coercion? This problem is often referred to as a second-order collective action dilemma. Coercion is a tempting solution to the problem of free riders, but it simply creates another free rider problem all its own.

In societies without states, that is often where the problem begins and ends. Collective action may occur, but its scale is usually small, and coercion plays little role in maintaining it. Where there is a state, on the other hand, coercion becomes a viable option for solving the collective action dilemma. And of course Olson was thinking about societies like ours—societies with states—not ones without states. States, as monopolists of the legitimate use of coercion within a particular territory, can simply force people to contribute to the collective action. As we all know, governments provide many goods and services to the general public: roads, drinking water, sewage service, national security. Governments are able to do this by spending tax money, and they get that money by forcing individuals and businesses to give it to them. Governments punish those who do not pay their taxes; contributing to this collective good is not optional. Military protection of a country's borders is a public good, but provision of that good is seldom a problem because contributing to the upkeep of the military is required by law. Other things provided by governments may not fit the definition of public good quite so well, but public-good-style arguments are often used to justify the use of the government's ability to coerce people to ensure that they are provided.[12]

In many cases, however, collective goods are created without overt coercion by the state or by anyone else. One important variable is the size of the group, and three of the solutions to the collective action dilemma that Olson provided are related to size. In all three, he argued that "small" groups, that is, groups with few potential members, have the advantage when it comes to organizing. It's important to remember that by "small," Olson did not usually mean the *actual* size of the group's membership, but the *potential* size of the group. That is, an environmental interest group that fights for cleaner air is a very large group by Olson's assessment, regardless of whether it has three dues-paying members or three million dues-paying members. Because all people who breathe would benefit from cleaner air, an environmental group working to provide cleaner air has an enormous number of potential members. The number of active members the group actually has on its membership rolls is an indicator of how successful it has been in organizing but does not change its underlying character as a "large" or "small" group under Olson's logic. It is certainly likely that an environmental group with many members will be more successful politically than an environmental group that has failed to attract many members, but this is irrelevant to Olson's point. For Olson, the environmental group would always be a "large" group—regardless of how many actual members it has—because the *potential* number of people who would benefit from its success is very large. Olson argued that such groups, like ZPG, would have a difficult time mobilizing and would garner memberships that are much smaller than the number of people who would potentially benefit from the organization's actions. These nonmembers are free riders.

Olson suggested that actors who have very large stakes in the issue at hand will have an incentive to act for the common good even if doing so ends up benefitting free riders. A large stake makes it more likely that the actor's eventual payback will exceed the effort expended—the benefits of participating will be greater than the costs. This is the case, for example, when candy manufacturers or soda bottlers lobby against tariffs and quotas in efforts to keep sugar prices low. Hershey Co., for example, has an enormous stake in whether wholesale sugar prices are 30 cents a pound or 36 cents a pound. A few cents a pound could cost the company millions of dollars. Spending a few hundred thousand dollars on salaries for lobbyists, who might be able to convince government officials to ease limits on sugar imports, would be money well spent. But for the average

BOX 3.2. TYPES OF GOODS

		Subtractability	
		Low	High
Excludability	Difficult	Public goods	Common-pool resources
	Easy	Toll or club goods	Private goods

Figure 3.1. A Typology of Goods. *Source*: The figure is based on one in Ostrom et al. (1994:7).

Economists distinguish four types of goods based on two variables, *subtractability* and *excludability*.

Subtractability refers to how much one person's use of the good prevents someone else from using it. This is also referred to as *rivalry*. A good with low subtractability is nonrivalrous, and a good with high subtractability is rivalrous. Private goods have high subtractability. If you eat a banana, no one else can eat the same banana. Common pool resources also have high subtractability. If you graze your livestock on a common pasture, no one else's livestock can eat the grass that your livestock eat. Public goods and toll or club goods have low subtractability. Your ability to benefit from the government's provision of the public good of national defense does not detract from your neighbor's ability to enjoy that same good. Your ability to travel on a toll road does not prevent others from traveling on that same road. The phrase "jointness of supply" captures essentially the same idea. If the cost of a good does not depend on the number of people who share it (i.e., the number of people consuming the good can increase without its cost increasing), then it is said to have high jointness of supply.

Excludability refers to the difficulty of preventing people from consuming the good if they did not help pay for it. People who consume a good without helping to pay for it are referred to as *free riders*. Public goods and common pool resources are both typified by low excludability. It is difficult or costly to prevent people from using them even if they have not helped pay for them. You may pay your

Box 3.2 continues

Box 3.2 continued

taxes and reap the benefits of national defense, but so may your neighbor who pays no taxes. You may help maintain your community's irrigation canals, but your neighbor who puts little effort into canal maintenance may still be able to use the irrigation system to obtain water for his fields. Toll goods and private goods, on the other hand, are typified by easy excludability. If you don't pay for a shirt, you won't be allowed to take it out of the store. If you don't pay the toll, you don't get to drive on the turnpike. The phrase "jointness of consumption" captures essentially the same idea. If it is not easy to prevent free-riders from consuming a good, then it is said to have high jointness of consumption.

In the study of cooperation, excludability is a more important variable than subtractability. If it is easy to exclude free riders from enjoying a good, then, regardless of its subtractability, it is not difficult to explain why the good would be provided: there is a profit to be made. But, if it is difficult to exclude free riders, the collective action dilemma sets in. As a result, much of the scholarship on cooperation concerns the provision of public goods and the management of common pool resources.

A third important variable not captured by this schema is *jointness of production*: Is collective action required for a good to be produced, or can an individual do it alone? This is important to the study of cooperation because high jointness of production creates an additional obstacle to collective action. If an individual or small group can create a good, then, ceteris paribus, it is more likely to be provided than if it requires collective action by a large group of people.

Collective good is another common term in the literature. It is often used simply as a synonym for public good. It can also be used as a way to refer to both public goods and common pool resources simultaneously. Because both of those types of goods share the crucial and problematic characteristic of low excludability, it is often useful to be able to refer to them together.

The concept of "*a* public good" should not be confused with "*the* public good." A public good is a technical concept that applies to certain kinds of goods and not others. *The* public good is political rhetoric. Perhaps because a good's status as "public" can be used to justify government efforts to supply it, a common political tactic is to refer to the provision of other types of goods as being ways to further "*the* public good." Indeed, it may sometimes be the case that *the* public good is best served by government provision of what are, at least technically, private goods. Consider firefighting, for example. Technically, firefighting is a private good. It can be—and once was—provided by

Box 3.2 continues

Box 3.2 continued

private brigades paid by insurance companies to preserve the properties they had underwritten. But if fire breaks out on uninsured property, the negative externalities—burned adjacent buildings, neighborhoods, and entire towns—that result may be considerable. Those negative externalities make it easy to justify treating this private good as if it were a public one, and the coercive power of the state can be used to eliminate the free rider problem simply by forcing everyone to pay for the good.

The same sort of logic is used with government provision of two other private goods, education and health care. Technically, both are private goods. If consumers don't pay tuition, insurance premiums, or physician's fees, then, in principle, it is easy to prevent them from attending school or receiving medical care. Nevertheless, education is usually seen as a public good, and health care is increasingly seen as one, as well.[1] Like fire insurance, health insurance and education are goods that would be easy for individuals to forgo. Why buy health insurance when you are healthy? Why pay tuition when you can educate your children at home? However, if people fail to buy health insurance, the rest of us are left with a choice of either refusing them health care or allowing them to receive it and seeing our own health costs increase. If parents fail to send their children to school, people fear that we will not have the kind of citizenry necessary to have a thriving democracy and economy. Because many people find the negative externalities that arise when people fail to buy health insurance and send their children to school so unacceptable, those who fail to consume health insurance and education are perceived as free riders. This contributes to the perception that they are public goods. So, for example, when the government provides free public education, the public good is not the classroom instruction itself (which remains a private good) but rather the alleviation of the broadly experienced negative externalities.

It is also important to note that calling something a "good" does not necessarily mean that everyone will like and appreciate the outcome and find it "good." While economic usage assumes that these goods or items have value, no moral judgment is implied by the word. The word "good" may also cause confusion because of its association

---

[1] The case for health care as a public good is strengthened by the fact that there really is a public good aspect to some kinds of health care. Specifically, infectious diseases create a collective action dilemma because an individual can free ride on others' efforts to stop them. For example, one can hope to avoid the flu and other infectious diseases by counting on others to be vaccinated rather than by getting vaccinated oneself. Noninfectious diseases and injuries do not have this quality.

Box 3.2 continues

Box 3.2 *continued*

with actual, physical "goods," as in "goods and services." If you substitute "result" or "outcome" for "good" in such phrases as "public
good" and "collective good," you will get a better idea of what they
really mean. In cases of collective action, the public or collective good
may be a particular social outcome rather than a physical object. In
the case of the Population Connection, for example, the public good
being sought would be whatever improvement to the world that its
members imagine might result from limiting population growth.

consumer who spends a few hundred dollars a month at the grocery
store, a few cents here and there has a relatively minor effect on a grocery
store bill. While such price increases may be annoying, it would not be
worth the shopper's time to open up a governmental relations office in
Washington and begin contacting government officials. That would cost
far more than the few extra cents or even few extra dollars spent each
week at the market. Instead, shoppers as well as smaller corporations free
ride on the lobbying efforts of big corporations, a phenomenon Olson
called the "exploitation of the great by the small." There are relatively
few entities in the world that buy anywhere near the amount of sugar
that Hershey uses per year. Thus, Hershey and other big corporations like
it make up a "small" potential group of industrial sugar buyers. So Olson's first solution related to size was simply that in smaller groups it is
more likely that each member of the group will have a share of the benefits pie that is large enough to make its efforts worthwhile. Each member's share of the collective benefits is likely to outweigh its efforts to
organize. Olson called these small groups *privileged* groups.

Olson's second reason why small groups have advantages over large
ones was that the effects of an individual's contribution to the collective
good *decrease* with the potential size of the group. This is the efficacy
problem: if the potential group is large, then it is hard for one person or
one individual company to make a difference or to unilaterally provide
the benefit. I may value a litter-free park and I personally may be willing
to dump my trash in the trash can and even to pick up the trash of others
around me, but if I have just attended a concert with thousands of people
in attendance, my efforts alone cannot make the park clean. Unlike industrial sugar buyers and their lobbying efforts, I cannot unilaterally provide the benefit of a clean park, or even make much of a contribution on
my own. Likewise, I may value clean air, but my forbearance from buying
an SUV, taken alone, will not make a noticeable impact on the quality of
the air, not even in my own neighborhood. I lose the benefits of driving

an SUV—roominess, road clearance, traction in the snow—but have not succeeded in providing a noticeably cleaner planet. I have paid a cost but received no obvious benefit. Given the seeming futility of unilateral action and the temptation of other actors to free ride, the efficacy problem causes many large groups to break down before they form. If provision of the good is suboptimal because of free riding (perhaps some pieces of litter are picked up, but many still remain), the few people who try to continue the action may have such a heavy burden to carry that they become discouraged and end up dropping out as well. Olson called groups that face such problems *latent* groups and expected that they would not organize naturally without additional intervention.

Olson's third solution related to size stems from the fact that it is easier to coordinate and monitor a small group. The costs of organization increase with size. This is a simple concept to grasp. If a member of a small group stops pulling his or her weight, other members of the group are likely to notice and apply social pressure (this may be as straightforward as a frown of disapproval). But in a large group, especially a very large group such as all supporters of clean air or all supporters of world peace, it is hard to know what all of the others in the group are doing. It is possible for large groups to begin to artificially overcome this monitoring problem by organizing into local suborganizations or cells, but of course the organization must be organized in the first place if it is to do this.

Despite his skepticism, Olson did see a way for large groups to get their acts together. Unless they are able to use coercion to increase people's costs for not cooperating, they have only one choice: increase people's benefits for cooperating. These private incentives are called "selective" benefits to distinguish them from the public good being sought.[13] Selective benefits, unlike public goods, cannot be attained by free riding. A selective benefit goes only to the person who joins, participates, or pays. It goes to members, not nonmembers. When political scientists join the American Political Science Association, they automatically receive three journal subscriptions, access to a site that posts information about job openings, and the ability to register for the annual conference, where they would meet other political scientists. If they did not join and did not pay their membership fees, they would not receive these bonuses. These bonuses are all selective benefits. The APSA provides collective goods to political scientists as well. It lobbies Congress on issues related to funding for political science research and provides general information about the profession on its website. Political scientists receive the benefits of these actions whether or not they choose to join the APSA. If the APSA succeeds in convincing Congress to allocate more money for political science research, any political scientist may apply for those grants, not only members of the APSA. It is easy to free ride on such collective benefits, and if these were the only benefits that APSA provided, it would have many fewer members than the fifteen thousand political scientists who pay their annual dues of up to $314. But in fact most political scientists do join because most of what APSA does is provide profession-related services to its members. The selective benefits provide enough incentive to prevent potential members from free riding, and thus APSA is maintained as a successful organization. This is a common structure for membership groups—not only professional associations like APSA and the American Medical Association, but trade associations like the Alliance of Automobile Manufacturers and the American Sugar Alliance, and issue advocacy groups like the Sierra Club, the National Rifle Association, and AARP all provide some level of selective benefits to their members.[14] Whether it is a magazine subscription, a coffee mug, or access to a conference, the vast majority of successful large membership groups provide material selective incentives to their members. If you want to go to the conference, you have to join the group. If you want to receive the magazine or get the discounted insurance, you have to join the group. The AARP has more than forty million members, making it the largest citizen membership group in the United States and one of a very few citizen groups that spends more than $1 million a year on lobbying. This is possible because it also is one of the best providers of material selective incentives among citizen groups, offering discounted insurance, travel, and other benefits, while charging only $16 per year to be a member.[15]

Selective benefits provide a way for groups to form that avoids the problem of free riding because there are no public goods for free riders to enjoy. However, once a group exists, it is then in a position to provide collective or public goods as well. A widget-manufacturing corporation, for example, is initially formed to create widgets for its customers, profits for its shareholders, and salaries for its employees—all selective benefits. It does not form in order to lobby the government for protective tariffs for the widget industry, and yet the fact that it exists gives it the wherewithal to do exactly that. Similarly, a trade association may form to provide networking and informational services to members of an industry, with those benefits being available only to paying members, but it may subsequently choose to lobby the government, thus providing a benefit to nonmembers, as well. Because in this scenario collective goods are provided as a by-product of the selective benefits provided by the group, this is known as the by-product theory of public goods provisioning, which parallels the biological theory of by-product mutualism discussed in the previous chapter.

By-product theory is one of the reasons why there are many more businesses and professional groups lobbying in Washington than there are issue-based groups without an occupational basis. Trade associations, professional associations, and corporations all form for reasons other than lobbying for collective benefits. They form to provide services to members of the profession or to manufacture some product or provide some service. But as a *by-product* of forming the organization to benefit members or stockholders, the organization may be able to shift a small portion of its income toward seeking collective goods—i.e., lobbying. Because issue-based groups form solely to provide the collective good of lobbying, they are much rarer. Counts of lobbying organizations in Washington, for example, show that 80 percent or more of all such groups are businesses, trade associations, professional associations, and other organizations whose members join because of their jobs.[16] Even so, by-product theory has helped some nonoccupational groups as well. The first mission of the National Rifle Association, for example, was to provide gun training for its members; only later did it begin to use some of its extra resources for the collective benefit of lobbying.[17]

Others' solutions: Solidary benefits, expressive benefits, patrons, and entrepreneurs

According to the Web of Science's citation database, *The Logic of Collective Action* has been cited in scholarly journals more than six thousand times. This overwhelming response reflects the fact that Olson had expressed what many had observed but none had presented so starkly and convincingly. As an economist, Olson had laid out his argument in mathematical form. This made it relatively easy for other formal modelers to build additions, extensions, and corrections to his model. But because Olson's theory was supported by logic and anecdotes, not systematic empirical testing, it also opened the door to a series of articles and books that subjected the theory to the real world, providing work for a large number of scholars and their students. Political scientists and sociologists found many cases in which groups had sometimes formed seemingly in spite of Olson's logic. Psychologists and economists explored the world of collective action in their laboratories. Anthropologists looked at how small-scale and stateless societies cope with free riders. The result is an enormous body of scholarship that we cannot hope to fully cover.[18] Instead, we will describe some major extensions of and challenges to Olson's path-breaking model.

Some of the most important work in this tradition actually predates Olson's book. In 1961, political scientists Peter B. Clark and James Q. Wilson described ways in which individuals might be persuaded to join a group.[19] Clark and Wilson identified three different types of incentives—material, solidary, and purposive—which would accrue only to individuals who participated in achieving the group's collective goal and thus

were not vulnerable to free riding. While material benefits, such as maga-
zine subscriptions for members only or a chance for discounted insurance, are tangible, solidary benefits result from the pleasurable interaction with other humans that comes from participating in a group. An
individual experiences purposive benefits when the group achieves its
purpose. Purposive benefits are also often referred to as "expressive benefits" because of the pleasure one gets from expressing oneself by working with others to further a cause.[20]

Olson accepted the idea that material selective benefits might help
groups form, but he parted company with Clark and Wilson over solidary and purposive or expressive benefits. He pointed out that it is possible to imagine a wide variety of other nonmaterial benefits, including
"erotic incentives, psychological incentives, moral incentives, and so on."
Because he saw no way to measure the degree to which individuals are
actually motivated by nonmaterial incentives, Olson feared that any attention to them would make his model untestable.[21] In Olson's view, all
such incentives also lead to tautology. If, for example, we say that people
join groups because they receive a "solidary benefit" by virtue of the fact
that they like to join groups, have we really explained anything? No, we
have simply said that "groups form because people like to form groups."
Olson, an economist, argued that the only way out of this tautology was
to explain the behavior of individuals solely in terms of the material selective benefits they would thus enjoy.

Material benefits do indeed help us explain many cases of group membership, even among groups that have organized primarily for seemingly
nonmaterial purposes. Sociologist David Knoke's survey of nearly nine
thousand members of professional, recreational, and women's organizations asked about selective benefits that members of such organizations
receive in the forms of such varied things as access to research and group
health insurance. Nearly 60 percent said they would drop out of the
group if they no longer received services in exchange for their dues. Modern membership groups are one thing, but even social movements and
rebellions have been shown to make important use of selective material
benefits. When Cesar Chavez was first trying to organize farm workers in
the early 1960s, the nascent union attracted many of its members by
promising benefits like burial insurance, a credit union, and help dealing
with government agencies. Material benefits for members were the main
attraction; public goods would come later. Of course, the organizational
efforts would have to succeed for the members to get these selective benefits, but at least the ability to exclude free riders was no longer a problem.[22] Even preindustrial peasant rebellions can be explained in part
using selective material benefits, according to political scientist Mark Lichbach. Food riots in the eighteenth and early nineteenth centuries often

included a chance to plunder food, while land riots sometimes seized common areas or involved the destruction of documentation of debts that the peasants owed.[23]

But the social and political world is filled with organizations full of members who do not seem to benefit materially from membership. Many issue-oriented groups do not offer any material selective benefits. Rather than treating these joiners as having made "deviant" choices, political scientists and sociologists set out to explain why individuals chose to join these noneconomic groups that espoused collective goals like clean air and zero population growth. Many of these explanations relied heavily on the solidary and purposive aspects of Clark and Wilson's theory. Political scientist Terry Moe called for a "broader view" of group membership in a widely cited journal article and recalculated Olson's model in another, allowing for imperfect information and adding variables for efficacy and selective benefits.[24] There were other prominent recalculations of Olson's model as well.[25] Dennis Chong's study of the civil rights movement argued that some of the most important selective incentives were linked to reputational concerns. These solidary or social incentives could be positive—those who participated and sacrificed in marches, sit-ins, and other protests were esteemed in the community—or they could be negative, as those who would not help were shamed. Chong told the story of a voter registration drive in Selma, Alabama, in 1965. At the time, being a school teacher was one of the best jobs an educated African American person in the South could hold, and throughout the South it was well known that teachers thus seldom participated in the protests—they had too much to lose. A local civil rights leader, the Rev. Frederick D. Reese, however, called the teachers to task and accused them of "relinquishing their citizenship by not registering to vote."[26] The teachers thus shamed decided to conduct their own march, and decided all would be required to attend unless they were elderly or disabled. If there were young children at home and both parents were teachers, only one had to attend. The organizers of the march took attendance so that nonparticipants could be identified. The only retribution for nonparticipation was embarrassment, both before their leaders and before their peers, whereas the punishment for participation might be the loss of their jobs. And yet turnout was nearly universal. "It is as if a comfortable myth shared by the teachers that their behavior was defensible was suddenly dispelled; now they would have to be either responsible citizens or not-so-admirable free riders. Therefore sometimes people shame each other into participating."[27] Such reciprocal reputational pressures were common throughout the movement, according to Chong.

Political scientist Robert Salisbury built on the solution to the collective action dilemma provided by selective incentives, but he also shifted the perspective. Selective benefits are indeed important, he argued, but who is offering those payments? Salisbury introduced the idea of the "interest group entrepreneur," and in so doing suggested another solution to Olson's dilemma. Olson's initial model assumed that for an interest group to form, each individual would contribute equally. Groups essentially would arise spontaneously, much as they had in previous theories, where the problems of initial group formation and coordination were never given much attention. Olson recognized those problems but saw few solutions to them, and because he is correct that groups with large numbers of potential members are extremely unlikely to arise spontaneously, he arrived at his pessimistic conclusions. Salisbury, however, being a political scientist, noted that groups do not arise spontaneously. They have leaders. Olson had noted the importance of selective material benefits to the organizational process but without considering who exactly was arranging for those material benefits to be provided. Enter Salisbury.

Salisbury borrowed the idea of the entrepreneur from economics and applied it to the world of organizing. The entrepreneur, in this model, is the group organizer or leader. An entrepreneur creates a product and then seeks to sell it to customers. A leader creates an organization and then seeks to attract members to join it. The leader builds in selective benefits for potential group members and makes it easy for those members to become a part. The members must contribute relatively little to the group, but because there are many of them, the collective benefit is likely to be provided. What does the entrepreneur get out of all of this? Selective benefits of his own. The dues of the members may pay his salary. The prestige of being the leader of such an organization may further his career or his social life. And in terms of purposive or expressive benefits, the organizer is running the show and guiding the organization, so the collective actions of the group are likely to best reflect his beliefs rather than the beliefs of the rank and file. While free riders may get benefits they don't pay for, they also don't have any voice in the nature of those benefits or how they are provided. In short, free riders can't be choosers. Members do have some voice in such things, but entrepreneurs have the most voice of all. Reading Olson, one would expect only "small" groups to form in this way, because only in small groups could it be expected that a single member would be efficacious enough and expect to receive a large enough portion of the collective benefit to make mobilizing worth his or her while. But once we recognize that selective benefits may differ from member to member and that the entrepreneur may receive very valuable selective benefits that differ from the selective benefits received by rank-and-file members, the problem is solved.

It is not an accident that citizen groups and social movements that remain active for more than a brief period virtually always have paid staff.

While these staff members may indeed care about the cause and may indeed be making less than they might make in the world of commerce, still they do receive a salary and often very friendly work environments. It is unlikely that many of them would remain and devote full-time hours to an organization if their paychecks ceased to be issued. Scholars who study social movements have even come up with an acronym to describe the bureaucratized aspects of social movements—SMOs, or social movement organizations—which often help direct the movement during its rise and certainly are what prevents the complete disintegration of the movement when passions cool.

Perhaps the most extreme example of Salisbury's entrepreneurial theory comes from a social movement in the early 1930s that advocated for stipends for the elderly, an effort that sociologist Edwin Amenta credited with laying the groundwork for modern-day Social Security. The effort began with a letter to the editor written by Dr. Francis E. Townsend in 1933, proposing that senior citizens who agreed to avoid working—freeing up a job for a younger person during the Great Depression—should receive a government payment of $150 a month. The proposal was radical, given that $150 was more than the median family monthly income at the time and given that the elderly were not yet an important political force.[28]

After writing the letter, Townsend worked with real estate agent Robert Earl Clements to form Old Age Revolving Pensions, Ltd., "to promote and secure by means of education and every other means, the adoption of the United States government . . . of plans and law providing for the pensioning of its citizens."[29] Townsend's brother Walter was added as the third member of the corporation. The organizers paid themselves $50 per week plus about the same amount in expenses, then set about to expand the movement. They created a newspaper and other Townsend Plan–related publications and hired two salesmen on a commission basis to sell the publications, collect donations, and organize branches of the club. Those salesmen also found other people willing to help start clubs—for a small commission, of course. Amenta compared the techniques used to those that Amway and Tupperware use to market household goods.

Why did the members themselves join? No doubt they supported Townsend's ideas and would have liked to see a pension plan for the elderly in place. But it is doubtful that they would have mobilized for this without the push from Townsend and his partners. For 25 cents in dues they essentially got to join a club that brought them together for social events. The cost to the members was relatively low; their effort was much lower than that of Townsend, his partners, and his salesmen. The members received solidary and purposive benefits, while the entrepreneurs reaped the material benefits of salaries and commissions. By the end of

1934 the movement boasted two thousand clubs and three hundred thousand club members across the nation (although Amenta thinks there were actually only about half this many).[30] The organization and its members galvanized public opinion behind the idea of senior pensions, and although it never achieved its particular goal, by 1939 a more modest version of elderly assistance had been passed by Congress in the form of amendments to the Social Security Act of 1935. These efforts to advocate for a collective good hardly arose spontaneously; the proposal would not have gone very far without the material selective incentives enjoyed by the entrepreneurs who began the organization.

Salisbury himself offered numerous similar (if somewhat less extreme) examples from U.S. history. He provided evidence showing that farm groups in the United States in the latter half of the nineteenth century did not arise spontaneously from the grassroots up, with groups of individuals simply deciding to cooperate. Rather, they were begun by enterprising individuals (including in several cases newspaper publishers who started the organizations to help build circulation) or with the help of government or existing organizations that subsidized the creation of the new organization. Entrepreneurs also often come from the older groups, and the older groups serve as a training ground—methods of organization were built upon rather than recreated from scratch.

The idea of the entrepreneur is closely linked to the idea of the "patron." While the entrepreneur helps overcome the collective action dilemma by doing more than his or her fair share to form the group—in exchange for selective benefits provided by the group once it is formed—patrons help overcome the collective action dilemma by paying more than their fair share. Both entrepreneurs and patrons have the ability to provide at least some of the collective good unilaterally (and can provide even more of the collective good if they attract followers), thus making even what Olson called a "large" group function more like a "small" or "privileged group."[31] Jack Walker's study of membership associations in the United States found that citizen groups working for public causes tended to get only about a third of their revenue from dues—that is, from rank-and-file members. Grants and large gifts from donors, on the other hand, made up more than 40 percent of annual revenues. These patrons were institutions that supported the cause—in the case of the grants—or wealthy individuals who supported the cause—in the case of the gifts.[32]

Even less institutionalized mobilizations often have an entrepreneur or patron behind the scenes. Lichbach cited examples from the late eighteenth century in which dissident elites whose upward mobility had been blocked incited peasant rebellion to use as a weapon. The incentives for the peasants to rebel were the side payments they received from the entrepreneur behind the scenes. Likewise, the political violence and ethnic

murders that followed the disputed December 2007 presidential election
in Kenya were at first attributed to spontaneous outpourings of rage by
supporters of the opposition party who were convinced that the election
had been stolen. More than one thousand people, including women and
children, were killed in the resulting violence, and hundreds of thousands
of people fled their homes. By the time the two parties had reached a
power-sharing agreement in late February, however, there was already
evidence that the violent mobs were far from unprompted. A UN report
found that while there were some spontaneous protests in cities and
towns, it was accompanied mostly by property damage, looting (a material selective incentive), and clashes with police. A separate type of violence occurred in rural areas throughout the Rift Valley, where vigilante
gangs targeted areas where opposition ethnic groups predominated and
set out to destroy property and kill any residents. These gangs were often
encouraged and organized by village elders, and the orchestration of
these activities is believed to have involved politicians at the highest levels
of government. In January 2012, the International Criminal Court or-
dered four prominent Kenyans, including two presidential candidates, to
stand trial on charges that they helped instigate the violence.[33]

It's clear what is in it for the entrepreneurs—political gain and in many
cases salaries or other material goods. If we limit ourselves to explanations in terms of rational Olsonian material benefits, then it is less clear
what motivates patrons. Some of the patrons are charities or government
agencies that give grants, so thinking about individual motivations is not
so important. In other cases it could be the prestige value of the donation.
Many of the local clubs started during the Townsend Movement were
funded by a single wealthy townsperson paying $25 to start the club
rather than one hundred members paying 25 cents each, Amenta's research showed. Such a gesture would help cement a reputation for local
importance. In yet other cases the patron may have ulterior motives, like
dissident elites who paid "preindustrial crowds" of peasants to riot in the
eighteenth and early nineteenth centuries.[34] And in other cases it may
simply be that the patron wants the public good so much for himself that
he decides to unilaterally provide it. To illustrate this point, Russell Hardin tells a hypothetical story of a Greek shipping magnate whose business
concerns prevent him from leaving his yacht, thus causing him to miss a
favorite opera. He pays to have the opera broadcast on radio so that he
can hear it—but of course in the process millions of other listeners get to
enjoy the opera as well. For the tycoon it is a private good that once provided can be shared by others.[35] According to Hardin, billionaire Howard Hughes actually did essentially the same thing. Frustrated by the fact
that his local television station went off the air at 11 p.m., leaving him
with nothing to watch in the days before VCRs and DVDs, he simply

bought the station and had it show movies all night long. In one sense it was a public good, because everyone in the broadcast area who was having a hard time sleeping got to enjoy the movies—their watching the movies did not subtract from anyone else's enjoyment of the movies. But for Hughes, the movies were a private good that he wanted and obtained for himself. The benefit to others was an unintended by-product.

The Olsonian perspective and selective incentives in the form of material goods and solidary or social benefits go a long way toward explaining group membership and participation. And yet, the idea of purposive or expressive benefits keeps reentering the fray. These benefits—which many have noted are essentially psychological benefits, since people participate because it makes them feel better or feel better about themselves—are especially important in social movements and in political organizations that are long on higher values and short on material benefits. Lawrence Rothenberg's study of members of Common Cause, for example, found that most members reported purposive benefits like "keeps government honest and fair," in explaining why they had joined.[36] Among the more political organizations in sociologist David Knoke's large survey, the selective material benefits in the form of member services were somewhat less important and lobbying efforts—a public good—were of central importance to nearly a third of members.[37] Edward Muller and Karl-Dieter Opp's interviews with university students and activists in Germany and the United States found that those who had engaged in "aggressive" protests that were illegal and/or violent indicated that they were not motivated by any selective incentives but were motivated by political and/or collective motivations. Far from trying to free ride, the respondents who thought such protests might be costly were more likely to participate.[38]

It might be easy to dismiss the studies above as simply reflecting the result of survey respondents giving the socially acceptable answers to questions, and indeed that may explain some of the answers. But experimental work by sociologists Gerald Marwell and Ruth Ames came to similar conclusions. In the late 1970s and early 1980s, Marwell and Ames ran eleven related experiments among high school students, college students, and economics graduate students.[39] In the experiments, conducted primarily by mail and phone, subjects were given tokens redeemable for several dollars of real money and told that they could invest the money in a private good with a fixed return, or choose to invest in a public good that would have a better return if enough students in the group contributed. The more students contributed, the higher the rate of return. The public good paid its returns, however, to all students, regardless of whether they contributed to the good, so it would be possible to get the guaranteed private return *and* the public goods return if at least some others in the group did not free ride. A strict Olsonian prediction for this situation would hold that no one would contribute to the public good, that all would free ride for fear of being taken advantage of and losing their investments. Instead, on average 40 to 50 percent of all tokens were invested in the public good. Mean investments in the public good went up to 60 percent in small groups (usually the students were told there were eighty students in their group, but in one experiment students were told there were only four in their group). But mean investment went up to 84 percent when the result was a nondivisible public good that all would share (improvements for a dorm common area) rather than payments back to individuals. Even when the stakes were increased so that a free rider who took part in the public good could earn more than $30—quite a lot of money for a high school student in the late 1970s—about a third of all tokens were invested in the public good. The closest Olson's predictions about free riders came to coming true was among economic graduate students: only 20 percent of their tokens, on average, went into the public good. Marwell and Ames published the findings in an article titled, "Economists Free Ride, Does Anyone Else?" For the most part, the subjects in the experiments cooperated far more than Olson's logic would predict. This was despite the fact that they played the games anonymously by phone and mail and were not in the same room as the students with whom they were "cooperating." When asked to explain why they had contributed as they had done, they cited ideas about "fairness."[40] And Marwell and Ames's work is hardly isolated. Dozens of scholars in the decades that have followed have run and rerun variations on these games and repeatedly have found that while the provision of the public good may have been "suboptimal" in an economic sense, still people contributed to those goods far more than Olson's theory would have predicted.[41]

Findings such as these are problematic for many scholars—even those who have documented the importance of expressive or purposive benefits—because they do not fit easily into a rationalist perspective. Lichbach rebels against relabeling anything that motivates people as a purposive selective incentive because it "redescribes collective action with another vocabulary."[42] Or as David King and Jack Walker put it, "There is something wholly unsatisfying about resorting to purposive benefits to make group membership appear rational. Desires for purposive benefits *are* desires for collective goods."[43]

What works?

In today's world, most voluntary groups that work toward some social or political collective goal have been formed by some combination of the by-product theory, entrepreneurs, and patrons. And, as we have seen, membership organizations frequently bolster their numbers with the aid of selective benefits, material and otherwise. Olson was no doubt right about the difficulty of organizing large groups of people with relatively low levels of individual efficacy and relatively small individual stakes in the collective payoff—at least he was right as far as his assumptions held. Olson's basic mathematical model assumed that each individual contributed equally and each received an equal payoff that came only from the collective good. Olson predicted that most organizing would be "suboptimal"—that is, that only a small fraction of the total possible members would participate, as was the case with ZPG. But when those assumptions are relaxed, allowing for unequal contributions to and payoffs from a collective outcome, the picture changes. Olson also did not consider the fact that, in the today's world, "suboptimal" organizing is often enough for a group to achieve its goals. After all, many organizations with social and political goals need only to organize well enough to make a convincing case for government action, which enables them to use the coercive power of government to achieve their goals, rather than trying to achieve those goals themselves through community organizing.

For example, advocates for the disabled in the 1980s sought to make public places and spaces more accessible to people confined to wheelchairs. One way of accomplishing this would have been to raise funds nationwide for money to create cut-in ramps on sidewalks in cities and towns around the country, go to each city and town council to request permission to do so, and raise awareness by putting social pressure on businesses to make sure that their buildings were wheelchair accessible. This most likely would have occurred community by community, with some communities organizing successfully (mostly the more wealthy ones, because it is easier to fundraise there) and others failing to organize and provide the collective good. The provision of the good almost certainly would have been "suboptimal" from Olson's perspective. Some towns would have ramps and others would not, and it would take constant vigilance and continual remobilization on the part of activists to maintain accessibility. Or, the advocates could do what they actually did, which was to create advocacy groups such as Americans Disabled for Accessible Public Transit (ADAPT) and the National Organization on Disabilities (NOD), and appeal to the government. These efforts paid off in 1990 when Congress passed the Americans with Disabilities Act, which prohibited discrimination against individuals on the basis of their disability and led to increased accessibility of public transportation, public buildings, and "reasonable accommodations" by potential employers of the disabled. Organizations like these, formed using some combination of the methods described in this chapter—entrepreneurs creating groups partially for their own benefit and drawing in members at low cost with the help of selective incentives—become the new big fish in the policy arena. They can afford to provide the public good of lobbying unilaterally and rely on the coercive power of government to do the rest.[44]

Efficacy, inefficacy, and irrationality

Another source of frustration with Olson's approach is the conflict between its assumption of rationality and the often irrational behavior of actual humans. Behavioral economists for decades have used laboratory experiments to document the many ways in which people fail to match the image of Olson's "rational, self-interested individuals,"[45] and examples from the real word are also not hard to find. For our purposes, the most interesting examples of apparent irrationality are those that help people overcome the collective action dilemma.[46]

As we explained earlier in this chapter, one main source of Olson's skepticism regarding the ability of large groups to organize is "lack of efficacy": One individual cannot hope to have much of an effect on the production of the collective good. For the individual member or potential member, this lack of efficacy is demobilizing because no matter how much the individual cares, he or she can't make a difference if everyone else free rides. But how good are people at assessing their own efficacy? Political scientist Terry Moe examined this question by studying five Minnesota economic groups with at least fifty members: the Minnesota Farm Bureau, Minnesota Farmers Union, Minnesota Retail Federation, Minnesota-Dakotas Hardware Association, and Printing Industries of the Twin Cities. These organizations, which had between 89 and 35,000 members, would seem to be bad places to find people acting "irrationally" or altruistically given that all are economic groups formed to provide economic benefits to their members. Moe surveyed members and asked them about the reasons they joined the organizations and why they stayed members. As would be predicted by Olson's logic, most ranked the services they received from their organizations as their most important reason for joining. These selective benefits included information, insurance, seminars, and many other services. More than half of the groups' members said that they would drop out if these services were discontinued. So far, Olson's rational, self-interested individuals seem to slightly outnumber the "deviants" who would stay in the group even if they stopped providing selective benefits. However, more than half of the groups' members also reported that collective reasons ranging from the concrete (e.g., lobbying on behalf of their industries) to the abstract (e.g., a sense of responsibility or obligation to the group) were their *most* important incentives for joining. Although Olson's theory accurately predicts a lot of human behavior in this case, the proportion of seemingly anomalous choices is large.[47]

There may be several reasons for this pattern of responses. Some researchers would simply discount the survey research approach for questions such as this. Perhaps people do not have enough accurate self-knowledge to say why they "really" joined the group, or perhaps they are just giving what they think are socially acceptable responses. Certainly we ideally would want to pair studies such as Moe's with other research designs to see if actual behaviors matched reported behaviors. But although there are many more groups that offer material selective benefits than those that do not, still there are numerous examples of groups that do not offer material selective incentives and yet manage to attract members. Are these just people who are making deviant choices, as Olson said of the ZPG members, or is something more going on?

Moe also asked a series of questions about individual efficacy. How important is the individual member's decision to join and stay with the group to the group's success? His survey included a question about what effect the respondent thought their individual (or individual company's) *own dues* had on the success or failure of the association's lobbying goals: a big effect, some effect, or no effect. Moe also asked about individuals' feelings of responsibility or obligation to remain a member. Two-thirds of the members saw their contributions as having at least some effect on the organization, and 8 percent thought they had a "big" effect. Interesting, the delusion of efficacy went up with group size. The Farmers Union, which had the most members, had 12 percent of its members claim that their dues had a big effect, while the figure for the printers' organization, which had the fewest members, was just 3 percent. This is just opposite of what Olson's model would predict, as individual efficacy should decrease with the size of the group. If people's estimations of their efficacy were inaccurate simply because of random errors in their rational decision making, then underestimation and overestimation should be equally common throughout the sample of members.[48] But Moe finds that this is not the case. Respondents who thought that their own dues made a big difference were more likely to have reported that they joined the organization mostly for collective, political reasons than for selective material reasons. Moe uses these findings to question Olson's assumption that individuals are rational decision makers. We are left with the somewhat unsatisfying conclusion that all of these people are simply in error and that groups form because people make mistakes about their efficacy.

To move forward, let's take a step back

It is clear that there is no more of a consensus among social scientists nowadays regarding how best to predict when collective action will and will not occur than there was when Olson's book was first published. Some follow in Olson's own tradition, emphasizing the ways in which even rational self-interested individuals can find it worth their while to stop free riding and pitch in. Others relax Olson's strict assumptions regarding what it means to be rational and self-interested by including non-material benefits as well as material ones. Others point out that the very idea of individual rationality is difficult to maintain in the face of evidence to the contrary. Still others focus on the ways in which history, culture, and institutions shape the worlds in which individuals operate. Although all of these approaches have merit, they cannot easily be reconciled within the existing framework of the social sciences.

What the social sciences do offer, however, is considerable. First, social scientists have generated a large body of empirical studies of cooperation in the real world, providing insights into what works and what doesn't when people try to overcome the collective action dilemma. Second, the social sciences have generated a wide range of theories and concepts that, although they may not yet cohere into a unified theory of cooperation, are still very useful tools for anyone who wants to understand, foster, or discourage it. Third, decades of work on collective action have given us a good understanding of the circumstances in which our various theories seem to work and those in which they seem to fail. Finally, social scientists, together with scholars in the humanities, have documented in great detail the historical, cultural, and institutional worlds in which we all live and which, through our actions, we continuously create.

Can the social sciences' different approaches to cooperation be reconciled with one another? We believe that they can, but to do that we need to take a closer look at cooperation's raw material: the individual.

CHAPTER 4

Cooperation and the Individual

From the proximate to the distal

Every approach in the social sciences begins with some theory of the individual. The approaches to collective action described in chapter 3 are no exception to this rule. However, most such theories attempt to explain behavior at only the proximate, mechanistic level. Consider, for example, Peter Clark and James Q. Wilson's idea that people join groups that supply public goods in order to experience "solidary benefits." In other words, people join groups because it makes them feel good to join with like-minded people and accomplish something they feel is important. Of course, this is true enough. But to say that solidary benefits explain the existence of groups is simply to say that people join groups because they like to. At best, this is simplistic. At worst, it is circular (as Olson himself pointed out). We can escape from this circularity if we remember the lessons from chapter 2 about levels of explanation. Clark and Wilson answered the question of what motivates individuals to contribute to a public good at the proximate level, but they left the other levels—developmental, distal, and evolutionary—unexplored. The solution is a dose of evolutionary thinking. The evolutionary approach to behavior recognizes the importance of proximate mechanisms but also asks how they came into existence in the first place. In this way, explanations provided by evolutionary scientists and those provided by social scientists can be recognized and appreciated for what they usually are: complementary to one another, not competing. In this chapter, we explore the ways in which evolutionary scientists are shedding light on the evolutionary roots of the proximate psychological mechanisms that underlie cooperation.

Skeptics will wonder—quite rightly—whether we are justified in making claims about the evolutionary foundations of proximate psychological mechanisms. After all, evolution acts ultimately on DNA, so if we want to claim that cooperation is influenced by evolution, then we would do well to find some evidence that human cooperativeness is grounded in biological processes that could conceivably be encoded in our genes. The idea that there really are specific biological mechanisms behind at least some aspects of cooperation is supported by recent work in behavior genetics. One common technique in behavior genetics is to compare identical twins, who are identical genetically as well as in appearance, to fraternal twins, who are no more similar genetically than regular full siblings. Identical twins are far more likely than fraternal twins to have similar rates of political participation.[1] Another study, using a different technique, found a relationship between voter turnout and two specific genes.[2] If such genes encode proximate mechanisms that influence behavior, then the brain may be the best place to look for them. Functional magnetic resonance imaging (fMRI), which reveals the areas in the brain in which oxygen use is greatest, is useful because it can show which parts of the brain are most active during cooperation (or anything else). And indeed, several fMRI studies have shown that certain parts of the brain are more active than others during cooperation.[3] Among the more relevant findings is that subjects who were more cooperative while playing experimental games show activity in areas of the brain associated with reward. This suggests that there are intrinsic benefits at the neurochemical level for behaving cooperatively.[4] In other words, just as Clark and Wilson claimed, it makes us feel good to work with others to achieve common goals.

Hormones provide another window onto the proximate psychological mechanisms that underlie cooperation. Oxytocin, the so-called "cuddle chemical," is best known for its role in milk letdown in nursing mothers, but it also has been associated in nonhumans with the ability to form normal social attachments more generally—certainly necessary for many forms of cooperation. A study by a group of economists and psychologists found that subjects who received nasal spray containing oxytocin were more generous in an experimental game than subjects who received a placebo. The authors of the study suggest that oxytocin "affects an individual's willingness to accept social risks arising through interpersonal interactions."[5] Oxytocin may also have a darker side, but still one that may be important for cooperation. A team of researchers from the Netherlands has shown that nasally administered oxytocin increased subjects' willingness to trust and cooperate with members of their own group, but it also decreased their concern for out-group members. In one study, subjects were presented with a famous moral dilemma involving a runaway trolley car. The trolley is headed toward a group of five people, but between the trolley and the group of people is a switch. If you don't hit the switch, all five will die. However, if you do hit the switch, the trolley will head off in a new direction and kill one other innocent person, instead. Do you do nothing, or do you act in a way that causes the death of one person but saves the lives of five others? The Dutch subjects of this experiment were more likely to sacrifice the one to save the many if the doomed individual was given a name indicative of an out-group, such as

Germans (e.g., Helmut) or Arabs (e.g., Ahmed) rather than a typical Dutch name (e.g., Maartens), but only if they had inhaled oxytocin rather than a placebo.[6]

However, neither genes nor hormones are destiny. We are only beginning to understand the complex relationships among genes, hormones, environments, and behavior. Consider, for example, a recent study of the effects of testosterone on cooperation conducted by members of the same team that looked at oxytocin's effects on play in economic games.[7] Female subjects in a double-blind experiment were given doses of either testosterone or a placebo underneath their tongues and then played the Ultimatum Game. Players who had received the testosterone made significantly higher offers than those who received the placebo. Crucially, the experimenters also asked the players whether they believed that they had been given testosterone or the placebo. Those who believed that they had received a dose of testosterone (and these beliefs were wrong about half the time) made significantly lower offers in the game. The researchers speculate that their subjects were acting on their folk theories of how testosterone makes people act. Clearly, our belief systems remain crucially important elements in understanding human behavior, and much work is left to be done on the proximate mechanisms that underlie cooperation.

Another approach to the causal chain that underlies cooperation is to come at it from the other end, that is, to imagine social situations that might have occurred among our ancestors and provided selection pressure in favor of a willingness and an ability to cooperate. This focus on ultimate or distal explanations is the approach that has dominated the evolutionary study of cooperation, and so it is the one that we will take in the rest of this chapter.

Reciprocity and the identification of cooperators

While we were writing the first draft of this book, a Stanford business professor named Robert Sutton published *The No Asshole Rule*, a slim, pithy volume that quickly reached the top of the best-seller lists. Sutton explained how to avoid and deal with the "bullies, creeps, jerks, weasels, tormentors, tyrants, serial slammers, despots [and] unconstrained egomaniacs" that can turn any workplace into a nightmare.[8] Sutton's book captured an important insight about cooperation: the first step toward successful cooperation is to avoid noncooperators and free riders. This is true not only of businesses and other large cooperative endeavors, but also of the most basic form of cooperation of all: reciprocity.

Box 4.1. The Reciprocity Bandwagon

A wide variety of social interactions have been called "reciprocity," but only a few of them involve actual reciprocity.[1] Here we provide a critique of some of the more popular ways in which he term has been used, misused, and abused.

Reciprocity: By itself, "reciprocity" is a perfectly good bit of everyday speech used to describe the back-and-forth, give-and-take quality of many social interactions. Reciprocity should not be used to refer to a single act in which a benefit is transferred from one party to another. Rather, to qualify as reciprocity, a social interaction must involve both the initial transfer and then the reciprocal response.

Delayed reciprocity: The addition of the word "delayed" serves to emphasize the fact that many reciprocal transactions involve a delay between the first gift or favor and the second, reciprocating one. Ethnographers have found that in many societies the delay between gifts or favors can serve as a way of strengthening the social bond between the two parties while also making it clear that the exchange is not the kind of immediate reciprocal exchange that typifies market transactions.[2]

Indirect reciprocity: Evolutionary biologist Richard Alexander coined this phrase to describe situations in which one party is helpful to another due not to a hope that the act will be reciprocated but rather because some third party may be favorably impressed by the act of kindness.[3] Unfortunately, "indirect reciprocity" does not actually fit the definition of reciprocity, as it is commonly understood. While we would like to see another term used for this kind of situation, this term is so often used and so well ensconced in the scientific literature that we expect it to remain for the foreseeable future. However, it is often possible to capture the same idea with such concepts as "audience effects" and "reputation effects."

Direct reciprocity: This phrase was coined in order to distinguish plain old reciprocity from "indirect reciprocity." Because it means the same thing as "reciprocity," we see it as unnecessary.

Network reciprocity: Evolutionary biologist Martin Nowak coined this phrase to describe a situation, discovered through his use of evo-

[1] We thank C. Athena Aktipis for discussions that led to this box.
[2] Cronk 1989.
[3] Alexander 1987.

Box 4.1 continues

Box 4.1 continued

lutionary graph theory, in which selection can favor unconditional altruists (whom Nowak calls "cooperators") if they occur in clusters.[4] Although the theoretical insights that Nowak and his colleagues have provided through their use of evolutionary graph theory are considerable, "network reciprocity" is an unfortunate phrase. In addition to following in the tradition among evolutionary biologists that we criticized in chapter 1 of confusing altruism for cooperation, "network reciprocity" is simply not reciprocity.

Strong reciprocity: According to economist Herbert Gintis, "strong reciprocators" are "predisposed to cooperate with others and punish non-cooperators, even when this behavior cannot be justified in terms of self-interest, extended kinship, or reciprocal altruism."[5] Like many who have been influenced by the evolutionary biological literature on cooperation, Gintis uses the term "cooperate" when he means "behave altruistically." Because the actions described by "strong reciprocity" cannot, even by definition, be reciprocated, they are not examples of reciprocity.

Weak reciprocity: This term was recently coined by economist Francesco Guala to distinguish "strong reciprocity" from plain old reciprocity.[6] As long as "strong reciprocity" is avoided, then this is yet another unnecessary term.

Generalized reciprocity: Anthropologist Marshall Sahlins used this phrase to refer to "transactions that are putatively altruistic," such as acts of sharing, hospitality, free gifts, help, and generosity.[7] "Generalized reciprocity" thus describes acts that are, by definition, not reciprocal. Depending on circumstances, "sharing" or "altruism" may be appropriate substitutes. This term appears to have been independently reinvented more recently by researchers working in the evolutionary tradition who use it to refer to a tendency to be altruistic toward other individuals in general after others have been altruistic toward you.[8] Although it is surely true that kindness can breed more kindness, this, too, is not reciprocity.

Negative reciprocity: Sahlins identified negative reciprocity as "the attempt to get something for nothing with impunity." In this category

[4] Nowak 2006.
[5] Gintis 2000:169.
[6] Guala 2011.
[7] Sahlins 1965.
[8] Pfeiffer et al. 2005.

Box 4.1 continues

Box 4.1 continued

he included haggling, barter, gambling, chicanery, theft, "and other varieties of seizure." Sahlins succeeded only in muddying the terminological waters. By including theft and seizure in the category, he used the term "reciprocity" to refer to things that are clearly not reciprocal. By including haggling and barter in the same category as theft, Sahlins implied that simple reciprocal market transactions are a form of exploitation. Sahlins's motivation thus appears to have been more political than scholarly.

Balanced reciprocity: Like "direct reciprocity," this is simply a new name for reciprocity, plain and simple. Sahlins introduced this phrase in order to distinguish actual reciprocity from "generalized" and "negative" reciprocity. It is unnecessary.

Reciprocal altruism: Evolutionary biologist Robert Trivers coined this phrase when he introduced reciprocity to the study of animal behavior.[9] Unfortunately, "reciprocal altruism" is a problematic term. By creating the impression that even so simple a form of cooperation as reciprocity involves altruism, it has contributed to the confusion about the meanings of "altruism" and "cooperation" that we criticized in chapter 1. Furthermore, "reciprocal altruism" is an oxymoron. If an altruistic act is reciprocated, then it is not actually altruistic because the actor receives direct fitness benefits. If it remains altruistic because it is never reciprocated, then it is not reciprocal, and selection would act against it. "Reciprocal altruism" thus describes something that selection would never favor.[10] Not surprisingly, evidence from studies of nonhumans indicates that selection never has favored reciprocal altruism. While reciprocity can be found among nonhumans, all purported cases of "reciprocal altruism" among nonhumans may be attributable to other evolutionary mechanisms, such as kin selection and mating effort.[11]

[9] Trivers 1971.
[10] Hamilton 1996:263; West et al. 2007.
[11] Clutton-Brock 2009.

According to game theorist Ken Binmore, reciprocity was discovered in a manner much like the Americas: it kept happening until everyone finally realized that it had happened.[9] As with so many things in the study of social behavior, David Hume had the basic idea back in 1740: "I learn to do service to another, without bearing him any real kindness, because I foresee, that he will return my service in expectation of another of the same kind, and in order to maintain the same correspondence of good offices with me and others. And accordingly, after I have serv'd him and he is in possession of the advantage arising from my action, he is induc'd to perform his part, as foreseeing the consequences of his refusal."[10] When ethnographers discovered that the world's societies contain a wide variety of reciprocal gift-giving systems, some simple and others quite elaborate, their study became a mainstay of the new discipline of anthropology.[11] By the time game theorist and eventual Nobel Prize winner Robert Aumann formalized the idea in the 1950s, he thought that it was already so generally known that he called it not the "Aumann Theorem" but rather simply the "Folk Theorem."[12] A few years later, Robert Trivers rediscovered the idea on behalf of biology.[13] The Christopher Columbus of the story, according to Binmore, is political scientist Robert Axelrod. Realizing that reciprocity was already a well-established area of research in the social sciences, however, Axelrod did not claim to have discovered any new continents. Axelrod's great contribution was a book, *The Evolution of Cooperation*, that was so compelling and widely read that people in disciplines across the social and life sciences finally realized that they had all been studying the same thing.[14]

Reciprocity's contribution to the study of cooperation is well summarized by Robert Aumann: "[C]ooperation may be explained by the fact that the 'games people play'—i.e., the multiperson decision situations in which they are involved—are not one-time affairs, but are repeated over and over."[15] Repetition is the key. If individuals have a good chance of interacting with each other in the future, cooperation can develop because they can hold each other accountable for favors given and owed.

Another way to say "repetition" is "iteration," and iterated games have been a very important tool for understanding how reciprocity can evolve. The single most important game in the literature on this topic, for good or ill, is the iterated Prisoner's Dilemma. In a noniterated, one-round Prisoner's Dilemma, two players must choose between two strategies, one labeled "cooperate" and one "defect." Imagine two individuals who had collaborated in a crime and agreed not to talk to the police if they were ever caught. "Cooperating" means adhering to that agreement; "defecting" means violating the agreement by talking to the police. If they both cooperate, they both get moderate payoffs, that is, light jail terms. If they both defect, they both get low payoffs (i.e., long jail terms), but not the lowest payoffs possible (i.e., very long jail terms). The dilemma arises because of the payoffs they receive when one defects and the other cooperates. In that situation, the defector gets the highest payoff possible in the game (no time in jail) while the cooperator gets the lowest possible payoff (a very long jail term). It is easy to see that in a one-round game, the best strategy is to defect because it is the only way

BOX 4.2. THE PRISONER'S DILEMMA GAME

The Prisoner's Dilemma is a two-person collective action dilemma. Its basic structure is simple:[1]

		Column player	
		Cooperate	Defect
Row player	Cooperate	Reward for mutual cooperation (R), Reward for mutual cooperation (R)	Sucker's payoff (S), Temptation to defect (T)
	Defect	Temptation to defect (T), Sucker's payoff (S)	Punishment for mutual defection (P), Punishment for mutual defection (P)

Figure 4.1. Structure of the Prisoner's Dilemma

The row player's payoff is always presented first in each box. Players move simultaneously, unaware of the other player's choice. To qualify as a Prisoner's Dilemma, the Temptation to defect must be greater than the Reward for mutual cooperation, the Reward for mutual cooperation must be greater than the Punishment for mutual defection, and the Punishment for mutual defection must be greater than the Sucker's payoff (i.e., $T > R > P > S$).

One way to cast the game is in terms of the original scenario involving two criminals who have been arrested but who have made an agreement with each other not to talk to the police. "Cooperate" means not talking to the police; "defect" means talking to the police, in violation of their previous agreement not to do so. In this case, the payoffs are negative because they represent time spent in prison:[2]

[1] Axelrod 1997:16.
[2] Chong 1991:6.

Box 4.2 continues

Box 4.2 continued

		Column player	
		Cooperate (i.e., keep mum)	Defect (i.e., talk)
Row player	Cooperate (i.e., keep mum)	1 year in prison, 1 year in prison	10 years in prison, 0 years in prison
	Defect (i.e., talk)	0 years in prison, 10 years in prison	5 years in prison, 5 years in prison

Figure 4.2. The Prisoner's Dilemma with Prison Sentences

If these payoffs are cast as negative numbers, then T > R > P > S, and so it is indeed a Prisoner's Dilemma–type game (0 > –1 > –5 > –10).

However, most people find it difficult to think in terms of negative payoffs. An easier way to think about the game is in terms of positive payoffs, as in this example:[3]

		Column player	
		Cooperate	Defect
Row player	Cooperate	3,3	0,5
	Defect	5,0	1,1

Figure 4.3. The Prisoner's Dilemma with Positive Payoffs

[3] Axelrod 1997:16.

to avoid the lowest possible payoff and because both players know that the other will be tempted by the high payoff associated with defection when the other party cooperates. This is essentially a two-person collective action dilemma. The Prisoner's Dilemma would thus seem to be a bad way to model cooperation. Its value becomes apparent when it is played repeatedly. In an iterated Prisoner's Dilemma, cooperation can emerge as a successful strategy as long as the game is likely to continue. An iterated Prisoner's Dilemma game presents players not so much with a collective action dilemma as with an assurance problem: It's in both our best interests to remain with us in the "cooperate/cooperate" box, earning steady, moderate payoffs, for round after round, but how can we trust each other to do so?

The dynamics of the iterated Prisoner's Dilemma were explored as long ago as the 1950s, but it was made central to the study of cooperation when Axelrod invited game theorists to submit strategies for playing the Prisoner's Dilemma and then played them against each other.[16] Tit-for-Tat, a very simple strategy submitted by game theorist Anatol Rapoport, emerged as the winner. A Tit-for-Tat player first cooperates. Thereafter, Tit-for-Tat does whatever the other player did in the previous round. Like other high-scoring strategies, Tit-for-Tat is "nice," meaning that it is never the first to defect. Such strategies rarely get the high one-round scores associated with defection while the other player cooperates, but they rack up large scores when both parties cooperate for round after round. In addition to showing that cooperation can become common even in a universe populated only by selfish actors, this now famous finding provided a starting point for an enormous number of additional studies based on the Prisoner's Dilemma. Among the many important findings in that literature is the observation that Tit-for-Tat can be beaten if some of the assumptions built into Axelrod's tournament are relaxed. For example, Tit-for-Tat cannot deal well with errors, such as defecting when you really should have cooperated.[17] When mistakes are a problem, an alternative strategy called Generous Tit-for-Tat, which sometimes cooperates even when the other player has defected, can do better than plain Tit-for-Tat because it can correct mistakes and avoid the low scores associated with mutual defection. A third simple strategy, Win-Stay, Lose-Shift (also known as Pavlov), in which each player starts by cooperating and then changes from cooperation to defection or vice versa depending on whether it is doing well or doing poorly, also does well when there are errors.

In Tit-for-Tat and Win-Stay, Lose-Shift, players are stuck dealing with each other even if they are not enjoying themselves. But of course in the real world we usually get to pick our cooperative partners and stop interacting with people whom we find to be uncooperative. In Robert Sutton's terms, we can try to avoid the assholes. Thus, a simple way to make the iterated Prisoner's Dilemma more realistic is simply to add the possibility of movement through space to the scenario. Athena Aktipis did this with a modified Win-Stay, Lose-Shift strategy called Walk Away.[18] Walk Away is essentially Win-Stay, Lose-Move: a player who is dissatisfied with how things are going with another particular player can simply move away and find someone else with whom to interact. Walk Away beats Tit-for-

Tat and Win-Stay, Lose-Shift because unhappy players can avoid defectors and find other cooperators.

Walk Away highlights the fact that one of the most fundamental aids to cooperation is simply assortment, that is, enabling cooperative individuals to spend more time interacting with each other and less time interacting with cheaters, free riders, and other uncooperative individuals. One way to think of walking away from an uncooperative individual is as a form of punishment. It is well known that punishment can make cooperation more likely. However, many forms of punishment are costly to the punisher and so constitute a public good. Any model purporting to explain cooperation that assumes the existence of costly punishment is essentially assuming that the collective action dilemma has been solved, which amounts to assuming away one of the biggest obstacles to cooperation. Another way to say this is that costly punishment constitutes a second-order collective action dilemma. If you hit a norm violator, he might hit back. To imprison the person, you would need to build a jail. Both activities are costly, and it would be easier to leave them for someone else to do, leaving the norm violator unpunished. Punishing those who fail to punish is no solution because, if such punishment is costly, then it is simply a third-order collective action dilemma. One way out of this logical tailspin is to look for ways that individuals can punish noncooperators at no cost to themselves. Walk Away provides a hint: just walk away. Ostracism is a simple way to punish someone for being uncooperative that not only costs you relatively little but also yields benefits in the form of the time you save by not trying to work things out with someone who won't work with you.[19]

The success of the Walk Away strategy also brings home an important point about psychological adaptations to reciprocity. Given that people vary in their quality as cooperators and assuming that our ancestors had at least some limited choice regarding their cooperative partners, it follows that we should have mechanisms that help us choose good partners.[20] In a species with long-term pair bonds and biparental care of offspring, perhaps the most important such choice would be that of a mate. After all, we expect our long-term mates not only to have sex with us but also to help us care for our offspring and to be committed to our relationship with them. Cross-cultural research shows that many of the characteristics people look for in long-term mates are similar to those we would expect them to look for in any cooperative partner. Dependability, emotional stability, intelligence, and sociability all rank highly for both men and women when looking for a long-term mate, and of course these are also the kinds of things we look for in cooperative partners outside mating, as well.[21] It follows that we should have the ability to identify such characteristics in others, even from minimal cues, and research shows that we do. In one recent study, men were able to accurately assess women's sexual attitudes (and thus, perhaps, their risk of being cuckolded) simply by looking at photographs of women's expressionless faces.[22] The key was facial masculinity (e.g., length of the lower jaw), which is an indicator of both testosterone exposure during development and a variety of personality and behavioral tendencies, including aggressiveness and a reduced affinity for children.[23]

Our ability to choose good long-term mates is one aspect of a broader ability to choose good cooperative partners. This ability appears early. Preverbal infants will choose to play with a toy that has been depicted as helpful to other toys over one that has been depicted as hindering other toys in their efforts to reach a goal.[24] One good way to associate with more cooperative people is to avoid uncooperative ones. Accordingly, in laboratory experiments people are better at remembering faces of people whom they are led to believe are untrustworthy or uncooperative than those of people who are supposedly trustworthy and cooperative.[25] People also remember the faces of uncooperative people better than those of cooperative ones. Toshio Yamagishi and his colleagues first obtained photographs of people who had played a one-round Prisoner's Dilemma game, keeping track of which ones cooperated and which ones defected. They then used those photographs in a test of other subjects' memories. Even though the second group of subjects knew nothing about how the people in the photographs had played the game, they were better at recalling the faces of the defectors than those of the cooperators.[26] Other studies have shown that people can tell the difference between more and less altruistic people simply by watching them on videotapes. The key seems to be the fact that altruists more often display Duchenne smiles, which involve not only the mouth but also the muscles around the eye and which are known to be better indicators of actual feelings than the forced "say cheese!" smiles that involve only the muscles around the mouth.[27] Sometimes a potential partner's cooperativeness may be less important than his other qualities, such as the ability to win a fight, and people across cultures are able to make accurate assessments of men's physical strength simply by looking at photos of their faces or listening to their voices.[28] Selection may have originally favored this ability because it helped our ancestors size up potential opponents, but it could easily have been co-opted for cooperative partner choice. After all, when push comes to shove, it's best to be on the side with the best shovers.

Another way to tell whether someone else is cooperative is simply to observe whether they cooperate. Anthropologist Michael Price has shown that the Shuar, an isolated group of horticulturalists in the Ecuadorian Amazon, do exactly that.[29] Shuar routinely work in groups, accomplishing such important jobs as clearing fields, harvesting crops, and building houses. Shuar estimates of one another's work effort correspond closely to Price's more systematic observations of such effort. Shuar also pay attention to whether those who shirk do so because they have chosen to shirk or because something is preventing them from contributing to the group project, and their impressions in this regard are a good match for systematically recorded excused and unexcused absences from work parties. Not surprisingly, those who were perceived as working harder and who had fewer unexcused absences had better overall reputations. These results are almost certainly not peculiar to the Shuar. As Price pointed out, every successful scheme for community management of shared resources involves some system of monitoring the behaviors of the members of the scheme.[30]

Price also predicted that people's willingness to help with collective work projects should be greater if they think others are also likely to help. Although data from the Shuar are not available to test this theory, public goods experiments in laboratory settings do support it. When subjects are able to move from group to group, more cooperative people leave groups populated by less cooperative people and go in search of fellow cooperators.[31] Similarly, when subjects are given information about each others' contributions to past Public Goods Games and an opportunity to form new groups for future rounds of the game, highly cooperative people tend to form groups with one another, thus reducing free rider problems. The simple ability to re-form groups results in average earnings that are higher compared not only to when subjects are stuck in their initial groups but also compared to when they are able to pay to punish each other regardless of whether they can re-form groups. Although punishment does result in higher average contributions, its cost reduces average earnings.[32]

Reciprocity, culture, and the avoidance of cheaters

At the same time that we humans ought to be trying to identify cooperators, we also need to be avoiding free riders and other kinds of cheaters. Fortunately, we seem to be very good at figuring out whether people have violated social rules. In the words of evolutionary psychologists Leda Cosmides and John Tooby, people have a dedicated "cheater detection mechanism."[33] A cheater, in this body of work, is someone who violates a social rule that requires a person to pay a particular cost in order to get a particular benefit. Evidence for the existence of such a mechanism comes mainly from the Wason selection task, a logic problem of the if-p-then-q variety.[34] A research subject is presented with four cards and a rule regarding what is written on the cards such as, "If a card has a D on one side, it must have a 3 on the other side." The four cards read D, K, 3, and

7. Subjects know that if there is a letter on one side there will be a number on the other side, and vice versa, but they do not know whether the rule has been followed exactly. Their task is to identify the cards that they need to turn over in order to find out whether the rule has been followed. Try it—the answer is in an endnote.[35]

If you chose the wrong cards to turn over, don't feel bad. Only a small minority of people have minds that are skilled enough at abstract logical problems to get it right.

However, performance on the Wason selection task improves greatly when the exactly same logical problem is presented not as a logical puzzle but rather as a quest to find out whether people are violating social rules. Again, you can try it yourself. Imagine that the rule is "If you borrow my car, you must fill the tank" and the cards read "borrowed car," "did not borrow car," "full tank," and "empty tank." Easy, right? If the car was not borrowed, you don't care what's on the other side because the rules do not apply. Similarly, if the tank is full, you don't care if the car was borrowed because even if it was, the rule has been followed. So you turn over the "borrowed car" and "empty tank" cards, which are logically equivalent to the correct answers in the abstract version. Almost everyone gets this version right. This contrast is displayed not only by the typical subjects of psychological experiments (college students), but also among the Shiwiar, a very isolated people in the Ecuadorian Amazon who have no system of formal education.[36] Others have found that what counts as "cheating" depends on one's perspective: people are better at spotting cheaters who work against their own interests than they are at cheaters who are on their side.[37]

Cosmides, Tooby, and their colleagues argue that the cheater detection mechanism reflects past selection in favor not of a general ability to deal with logical problems but rather in favor of an ability to identify, and then perhaps avoid or even punish, individuals who violate social rules. More specifically, they identify the selection pressures behind the cheater detection mechanism as arising from social exchange, starting with simple reciprocity of the you-scratch-my-back-I'll-scratch-yours variety. Another possibility was recently suggested by Natalie and Joseph Henrich.[38] They argued that because such rules are unavoidably cultural, such an ability should be seen as a product of gene-culture coevolution, an idea we described briefly in chapter 2. This observation is based on the fact that "cheating" in this body of work is different from defecting in the Prisoner's Dilemma game and other kinds of noncooperation in that it presupposes the existence of a shared social rule, and therefore of culture. We think both views have merit. The situation might be clarified by a realization that the scenarios used in Cosmides and Tooby's experiments come in two broad varieties. In some of them, the benefit is a favor from someone else, such as borrowing her car, and the cost is doing a favor for her in return, such as leaving the car with a full fuel tank. That scenario is based on reciprocity, so our ability to think through it may indeed be grounded in selection pressures arising from reciprocal social exchange. In other scenarios, the cost is something more arbitrary, such as getting a painful tattoo, and the benefit is in the form of a privilege, such as the right to consume an aphrodisiac. Despite the fact that this kind of scenario involves no reciprocity, people are still good at identifying cheaters in them. It may be that an ability to notice when one's favors are not reciprocated, which would not require gene-culture coevolution, was coopted by gene-culture coevolution in favor of an ability to identify people who violate other kinds of social rules regarding benefits and associated costs.

In our view, reciprocity is an essential part of any explanation of human cooperation. However, we would be remiss if we did not recognize that some scholars are skeptical about this. For example, Robert Boyd and Peter J. Richerson have argued that reciprocity cannot explain the development of cooperation in large groups.[39] They came to this conclusion by assuming that cooperative acts must benefit everyone in a group, not just the two reciprocators. If that were truly necessary in order for selection to favor reciprocity, then reciprocity would indeed have a hard time explaining very much about human cooperation. But of course the theory of reciprocity includes no such requirement. For reciprocity to be the basis of widespread cooperation, it is only necessary for each individual in a reciprocating pair to benefit.[40] It does take two to tango, but, fortunately, two can tango just fine without anyone else benefitting from the performance.

Generosity as performance, part 1: Indirect reciprocity

> Human society would be impossible without the ability of each of us to know, individually, a variety of neighbors. We learn that Mr. X is a noble gentleman and that Mr. Y is a scoundrel. A moment of reflection should convince anyone that these relationships may have much to do with evolutionary success.
> George Williams 1966:93

Cooperative partner choice and cheater detection are good examples of social selection, the process introduced in chapter 2 that occurs when organisms generate selection pressure on members of their own species. This form of social selection must have helped shape our species in important ways. Clearly, if people can detect and associate with cooperators and detect and avoid cheaters, then it would behoove them to also have the ability to prove to others that they are cooperators and not cheaters. This is the basic idea between two closely related evolutionary approaches to cooperation: indirect reciprocity and hard-to-fake signals.

The idea behind indirect reciprocity was around long before it had a name or a formal model. Williams captured the gist of it in the quote above. Whereas plain old reciprocity involves just two individuals (say, Abe and Ben), indirect reciprocity involves a third (Charlie).[41] If Charlie is pleased by what he sees Abe do for Ben, he may treat Abe nicely in the future, perhaps seeking him out for some cooperative venture. If Charlie is displeased by Abe's behavior toward Ben, he may avoid Abe or perhaps even punish him in some way. Either way, he may tell someone else (Doris) about Abe's treatment of Ben. Even if Charlie himself has been treated well by Abe, he might think again about working with Abe in the future if Abe treats Ben badly. As humorist Dave Barry once noted, "A person who is nice to you, but rude to the waiter, is not a nice person."[42] While Williams may have been the first evolutionary biologist to recognize this possibility, it was Richard Alexander who gave it a name that stuck and who explored its relationship to human moral systems.[43] In recent years, however, Martin Nowak and his colleagues have been indirect reciprocity's main proponents, arguing not only that it is the key to the puzzle of human cooperation but, because of its importance to the evolution of social intelligence and language, that it is also the key to understanding our species' distinctiveness.[44]

Indirect reciprocity has been the subject of a large number of modeling and simulation studies. One early model imagined cooperative players being rewarded by others connected through looped chains and found that indirect reciprocity was likely to lead to cooperation only in small groups with tightly looped networks of players.[45] However, this model was missing a crucial real-world element: reputation. Including reputations turns out to be the key not only to indirect reciprocity in the real world but also to making it work in simulations. One way to provide agents in simulation models with reputations is through a method called "image scoring," an approach developed by Martin Nowak and Karl Sigmund. Despite the name, image scoring has nothing to do with actual visual images. In image scoring, individuals who give help to others lose resources but get higher image scores, while those that pass up such opportunities get lower image scores but keep their resources. A player will then give help (i.e., resources) to another depending in part on whether the recipient's image score is above a certain threshold value. Under a variety of different assumptions about the strategies players use and how they gain knowledge about other players' image scores, this leads to populations in which strategies that accept the short-term cost of helping another in exchange for the boost to their reputations—and hence to the amount of help they receive from others in the future—predominate.[46] This finding soon received support from a laboratory study on the effects of reputation in an experimental economic game. Claus Wedekind and Manfred Milinski set up a game in which individuals could choose to give each other money at some cost to themselves, with their generosity being made known to the other players.[47] As Nowak and Sigmund (and everyday experience) would predict, players who gave more also received more.[48]

Although Nowak and Sigmund referred to the helping that players do in this model as "cooperation," it does not fit the definition of cooperation given in chapter 1 because it is an act by an individual rather than two or more individuals working together. Nevertheless, their model was an important step in the evolutionary analysis of human cooperation, and it has inspired a large number of subsequent studies. One important development is a shift from image scoring to "standing," an idea inspired by Robert Sugden's discussion of "good standing."[49] While a player's image score merely reflects his history of donating or not donating, his standing reflects his intentions as well as his actions by distinguishing between justified failures to donate (i.e., when the recipient is of low standing) and unjustified ones (i.e., when the recipient is of high standing). This brings us closer to how reputations work among real people: helping a deserving person is an action worthy of praise, but helping a criminal or miscreant, even a needy one, is condemned. The analytical payoff of standing versus image scoring becomes apparent when there is a chance that players will make errors when they play. Olof Leimar and Peter Hammerstein showed that players who attend to standing usually outcompete those who use image scores, while Karthik Panchanathan and Robert Boyd demonstrated that when players make errors, cooperation is more stable when it is based on standing than when it is based on image scores.[50]

Later, Panchanathan and Boyd made a contribution to the literature on the collective action dilemma by showing how indirect reciprocity can eliminate the second-order free rider problem.[51] As we have seen in previous chapters, collective action is undermined not only by those who fail to contribute to the public good but also by those who fail to punish those who do so. This is the second-order free rider problem. Panchanathan and Boyd imagined a situation in which individuals first have the choice of contributing or not in a collective action game. Then, individuals play an indirect reciprocity game in which they follow one of three strategies. "Defectors" do not contribute in the collective action game and do not help anyone in the indirect reciprocity game. "Cooperators" contribute in the collective action game and help those who need help in the indirect reciprocity, though an error term included in the model ensures that they will sometimes fail to do so. "Shunners" contribute in the collective action game and never help anyone who is in bad standing in the indirect reciprocity game. Shunners also help needy individuals with good standing in the indirect reciprocity game, though, like Cooperators, they sometimes make mistakes. Failure to contribute in the collective action game is more damaging to a player's reputation than merely failing to help someone in the indirect reciprocity game. A population of Shunners can resist invasion by both Defectors and Cooperators, creating an incentive for contributing to the collective action. Because Shunners themselves benefit from not helping players in bad standing, there is no second-order free rider problem. This model was inspired by a laboratory study by Manfred Milinski and his colleagues in which players alternated between Public Goods Games and indirect reciprocity games, one after another. Stinginess by a player in the Public Goods Games resulted in lower donations to him by other players in the indirect reciprocity game, creating an incentive to contribute more to future Public Goods Games. Thus, indirect reciprocity helped solve the collective action dilemma.[52] From an empiricist's point of view, the appeal of both Panchanathan and Boyd's model and Milinski et al.'s laboratory study is that they resemble the real world, where reputations are not isolated within particular realms of behavior (e.g., collective action situations versus opportunities to help other individuals) but rather reflect individuals' behaviors across many types of social interactions.

For social scientists, who have long recognized reputation's crucial role in human cooperation,[53] evolutionary theory's discovery of this fact might seem to come a bit late. Indeed, everyone is familiar with the effect an audience, or the lack of one, has on human behavior. Consider, for example, the rudeness of drivers versus the politeness of pedestrians. The Internet has created such a large space for anonymous rudeness that it has led to a folk theory regarding anonymity's impact on behavior. The Greater Internet Fuckwad Theory "posits that the combination of a perfectly normal human being, total anonymity and an audience will result in a cesspit." In other words, when Internet users are anonymous, inhibition and empathy seem to go out the window. Some website managers have responded to this problem by requiring users to register under their own names, thus removing the veil of anonymity, before posting comments.[54]

What evolutionary science can contribute is its ability to identify specific psychological adaptations associated with reputational concerns. Several recent studies have made progress on this front. All demonstrate the important effect that an audience, even one that is only hinted at, can have on rates of cooperation. One of the most interesting was conducted using the simplest materials available: a coffee pot, a donation jar, and photographs of eyes. Like many offices, the Department of Psychology at the University of Newcastle maintains a common coffee pot alongside a donation jar and a sign asking for people to make voluntary contributions to help pay for the cost of the coffee, tea, and milk. Three members of the department set their colleagues up as unknowing experimental subjects by posting two different kinds of images on the wall above the pot. One week, there would be a photograph of flowers. The next week, there would be a photograph of a pair of eyes. This went on for ten weeks. The amount of tea and coffee consumed was estimated by keeping track of the amount of milk used. Every time the picture shifted from flowers to eyes, the amount in the jar at the end of the week increased. Every time the picture shifted from eyes to flowers, the amount decreased.[55] In a follow-up study conducted in a cafeteria, pictures of eyes and flowers were paired either with a sign admonishing people to clean up after themselves or one asking them to consume only food and drink purchased in the cafeteria. Regardless of which sign they were posted alongside, the eyes generated more table clearing than the flowers.[56]

More controlled studies of the same phenomenon have revealed a similar effect. An image of a robot on a computer screen increased donations in a Public Goods Game.[57] Stylized eyespots on a computer screen were enough to increase donations in a Dictator Game.[58] Even the most minimal hint of a face is enough to elicit this effect. Mary Rigdon and her colleagues divided their subjects into two groups and had them play a Dictator Game. Some were given a sheet on which to record how much, if any, of $10 to allocate to an anonymous subject that included three dots arranged to look vaguely like a face, with two on top and one on the bottom. This was inspired by an earlier finding that such a "face" is enough to stimulate the part of the brain responsible for face recognition. The rest of the subjects were given a decision sheet that was identical except that the "face" was upside down, with two dots on the bottom and one on the top. Even a depiction of a face this minimalistic was enough to make people more generous.[59] Just as a false sense that an audience is present can be stimulated very easily, so can a false sense of anonymity. Chen-Bo Zhong and his colleagues created an illusory sense of anonymity among their experimental subjects by simply dimming the lights in a room or having them wear sunglasses. Both led people to cheat more and to behave more selfishly than people in rooms with normal lighting and no sunglasses.[60]

These findings are significant for reasons that go beyond the study of indirect reciprocity and cooperation. A common critique of evolutionary psychology is that it is simply folk psychology with a Darwinian makeover. While it is indeed true that folk psychology and evolutionary psychology sometimes make similar predictions, they also frequently contradict each other. More often, though, evolutionary psychology addresses issues and makes predictions that folk psychology never considers. The effectiveness of even minimalistic depictions of faces in making people more generous is a case in point. We know of no folk psychological theories that would predict such a thing. What evolutionary psychology and evolutionary theory more broadly bring to the table is the ability to derive novel predictions from evolutionary theory and test them in a rigorous fashion.

Even the specter of a supernatural observer is enough to make people more generous. In an experiment using the Dictator Game, subjects who were primed with God concepts through a task involving the unscrambling of sentences gave considerably more than unprimed subjects. Interestingly, the effect was seen in both theist and atheist subjects.[61] Of course, actual audiences also have effects on behavior. In a study involving English high school students, average contributions to a public good increased (and, correspondingly, the retention of resources by individuals decreased) when everyone's actions were made known to each other, but not when privacy was maintained.[62] In two experiments where subjects were given a chance to pay to punish others who had committed moral violations, punishment increased when other subjects were to be told of their choice. Even when only the experimenter was aware of subjects' choices, more subjects punished and spent more to do so than when their choice was completely anonymous.[63]

Because reputations are usually spread through language, the theory of indirect reciprocity also gives language a leading role in the evolution of our species' high degree of cooperativeness.[64] Although the origins of language are beyond the scope of this volume, it does not take much imagination to see how even a rudimentary ability to share information about others' behavior could have provided our ancestors with a variety of benefits and encouraged the further development of our linguistic abilities as well as our ability to engage in indirect reciprocity. Today, gossip about others constitutes about two-thirds of what people talk about in casual everyday conversations.[65] The ability to talk about others when they are not actually there is an example of a broader ability found in human language but not in the signaling systems of nonhumans: displacement. While nonhuman signals make sense only in reference to things that are present ("Look out! It's a leopard!"),[66] humans can discuss things that are not there. Someone can tell you how to get from Penn Station to Grand Central even if you are currently sitting at home on your couch. Or, someone can tell you about his colleagues, neighbors, and relatives, spreading good and bad reputational information in the process.

Although language differs from nonhuman signaling systems in many ways besides just displacement, some linguists identify it as language's single most important innovation.[67] Given that displacement is essential to our ability to gossip about reputations and that such gossip is a key to the power of indirect reciprocity, it is easy to imagine that indirect reciprocity and language coevolved in a positive feedback loop. Because language is an aspect of culture, indirect reciprocity is thus best understood as a product of gene-culture coevolution.[68] We know from studies of other kinds of selection (e.g., the effects of female mate choice on male courtship displays) that social selection feedback loops of this kind may be very powerful, making indirect reciprocity indispensable for understanding our highly social and cooperative species' rapid trip down an unusual evolutionary pathway.

Although we are convinced that indirect reciprocity is a key to our species' high levels of cooperation, we must acknowledge the fact that others are more skeptical. For example, Natalie and Joseph Henrich have claimed that "a standard indirect reciprocity model cannot sustain significant cooperation in groups much larger than dyads or triads."[69] If that were true, then indirect reciprocity would be of little help in explaining our species' high levels of cooperation. However, Henrich and Henrich's argument against indirect reciprocity is similar to Boyd and Richerson's argument against reciprocity: They assume that the costs and benefits of cooperative and uncooperative acts must be felt by everyone in the group. Sometimes, this is indeed the case. When we work with a group and we decide to withhold cooperation from someone else in the group, we must simultaneously withhold it from other individuals in the group.[70] Certainly, this is a real problem that people face every day. But it does not mean that there is no work for the theory of indirect reciprocity to do in explaining cooperation beyond dyads and triads. People live, work, reciprocate, and cooperate in networks. Reputations gained in dyadic and triadic interactions spill over into group contexts, and vice versa. Henrich and Henrich have argued that people living in large populations will have trouble directing their aid and cooperation toward others with good reputations because it will be hard to spread and store all of the necessary information about individuals' reputations. In our view, this ignores humans' remarkable capacity, due in large part to language, for doing exactly that.[71] Indeed, this seems to us to be a very large part of human social life.

Generosity as performance, part 2: Hard-to-fake signals

It pays to have a reputation for such things as generosity, kindness, trustworthiness, a willingness to follow even arbitrary norms, a willingness to punish those who do not show these characteristics, and a willingness to share accurate information about others' social behaviors. But how can you make sure that people know that you have these characteristics, particularly when they might be skeptical about your claims? This is where signaling theory can play a role in our understanding of human cooperation. Not all signals are equally believable. The ones that are most likely to convince a skeptical receiver are those that are hard to fake, that is, signals that individuals without the qualities being advertised find difficult to pull off. Such signals are often referred to as "costly," and it is indeed true that one way to make a signal hard to fake is to make it costly in a way that only honest signalers can afford. However, because sometimes signals either are hard to fake for reasons other than their cost or are costly for reasons other than a need to be believable, we prefer to refer to them as "hard to fake."[72] Hard-to-fake signals that are not costly are called indexical signals, a bit of terminology borrowed from semiotics.[73] For example, tigers mark their territories by scratching as high on trees as they can.[74] This signals to other tigers not only that they have been there but also their size. A scratch high on a tree trunk is not a particularly costly signal, but, at least until tigers learn to stand on boxes, it is a hard one for a small tiger to fake. Signals can be costly without being hard to fake simply because they need to be loud or otherwise conspicuous in order to get through to a receiver in a noisy environment. Signaling theorists refer to this as "efficacy cost" to distinguish it from the "strategic costs" that make some signals hard to fake.[75] For example, people yell in crowded nightclubs not to show off their powerful vocal cords but simply to be heard above the background noise of amplified music and everybody else who is also yelling just to be heard.

The value of hard-to-fake signals was recognized independently by both biologists and economists. Although biologists are most interested in signals designed by natural selection, economists (and other social scientists) are concerned mostly with signals designed by humans (e.g., advertising agents). However, because the same principles of signal design apply to both, we can use the same body of theory to study signals regardless of how they were designed. In biology, the classic problem is mate choice, which very often involves females choosing among males. In such a situation, selection would favor males who find ways of displaying the high quality of their genes in ways that cannot be imitated by low-quality males. This results in the great variety of conspicuous and sometimes even dysfunctional characteristics that are often easiest to spot by comparing the males and females of a particular species—peacocks versus peahens, for example.[76]

In economics, Michael Spence's theory of job market signaling provides an example of a hard-to-fake signal that is both familiar and clear.[77] Firms would like to hire focused, hard-working people, but how can they distinguish them from the unfocused and lazy ones? Focused, hard-working people, for their part, want to make sure that prospective employers can tell the difference between them and their unfocused, lazy competitors. One way to show that you are capable of lots of focused, hard work is to gain admission to an elite university and earn honors while you are there. Thus, employers have a reason to attend to educational achievements even if they have nothing to do with the specific job for which they are hiring. Hard-to-fake workplace signaling may continue even after one has been hired. We all know some model workplace citizens, people who work late, volunteer to take on additional tasks, bring donuts to meetings, and so on. Sabrina Deutsch Salamon and Yuval Deutsch have suggested that those sorts of "organizational citizenship behaviors" may constitute a hard-to-fake signal of otherwise intangible qualities such as conscientiousness, thoughtfulness, and commitment to the organization.[78] Political scientist Ken Kollman has applied Spence's idea to the phenomenon of grassroots lobbying. Organizations that do have the support of large numbers of people find it easy to create grassroots lobbying campaigns. Others find such campaigns to be very expensive and are less likely to attempt them at all.[79]

Although it is often said that hard-to-fake signals arise when there is a conflict of interests between signalers and receivers, this is not quite right. Hard-to-fake signals are most useful when there is a broad conflict of interests between classes of signalers and receivers (e.g., males and females; job applicants and employers), but where particular signalers and particular receivers have at least a temporary confluence of interests. Peahens have an evolved skepticism toward peacocks' claims about their own quality as mates, but it is in their best interests to mate with peacocks who are of truly high quality. Employers have a learned skepticism regarding job applicants' claims about themselves, but it is in their best interests to hire applicants who are of truly high quality. The link between the signal and the underlying quality it displays makes it both informative and believable.[80] While these sorts of situations do involve conflicts of interest between broad classes of signalers and receivers, the confluences of interest between receivers and high-quality signalers, whether males or job applicants, means that they have much to do with coordination, as well. As we will see in chapter 6, coordination problems are solved through the creation of shared knowledge: Everyone needs to know the solution to the problem, and they also need to know that everyone knows it. Hard-to-fake signals create common knowledge by making it clear to receivers that, while they may have reason to suspect that their interests and those of the signalers don't coincide, in fact they do.

The process employers use to select new employees (and vice versa) is a form of social selection. Social selection creates incentives to send signals about one's quality as an ally in a variety of circumstances, and not all of them will lead to signals that one might call "prosocial." For instance, if you are looking for people to be on your side in an armed conflict, you might want to forego such characteristics as agreeableness and focus instead on aggressiveness. However, it is often the case that social signals do take prosocial forms, such as public generosity and participation in group defense.[81] One reason for this is broadcast efficiency.[82] If you want to get a message out to a lot of people, you need to get their attention. A simple way to do this is to do something that they will appreciate, such as provide a public good. Thus, public generosity may serve as a signal of one's status and ability to control resources or one's cooperative nature.[83] Such generosity is not cooperation as we have defined it in this book, but it is relevant to the study of cooperation more broadly. By making it worthwhile for an individual to provide a public good without help from others, such "competitive altruism," as evolutionary theorist Gilbert Roberts has called it, has the potential to solve the collective action dilemma simply by removing it. By creating a system of unambiguous signals of cooperativeness, it also has the potential to enhance the effects of reciprocity and indirect reciprocity.

One of the most intriguing examples of generosity as a signal to others is provided not by any human society but rather by a bird called the Arabian babbler. Amotz Zahavi and his colleagues have been studying babblers in an Israeli nature preserve since 1970.[84] Babblers are highly social creatures. They live in groups, and each group defends a territory against both neighboring groups and individual babblers who are not part of a territory-holding group. Babblers do lots of nice things for each other. When they eat, one of them will act as a sentry, on the lookout for predators, and not eat despite being hungry. Adult babblers give each other food and feed each other's offspring, again despite being hungry themselves. Rather than flying away when the group is under attack, they risk their own lives by attacking dangerous predators such as raptors and snakes. All of this is fascinating. But what is most intriguing about Arabian babblers is that they actually compete for the right to engage in these altruistic behaviors. For example, higher-ranking babblers expel lower-ranking subordinates from the sentry post, sometimes even shoving and hitting the more stubborn ones. Lower-ranking babblers also seek sentry duty, but rather than trying to expel a higher-ranking individual from the guard post, they position themselves nearby and wait for an opening. Keep in mind that sentries cannot eat, so their eagerness to engage in this behavior is truly an altruistic act. Babblers also compete with each other for the right to feed the group's young, and feeding among adult babblers generally goes only one way: from higher-ranking individuals to lower-ranking ones. Zahavi interprets this pattern in terms of signaling theory.

Babblers use altruism to signal their quality to each other. The payoff is social status, which, in turn, pays off in terms of reproductive success. Social status is a limited resource, and those who seek it are playing a zero-sum game. Such high stakes can lead to strong selection pressure, even in favor of helping one's competitors.

Like Arabian babblers, people also often compete for opportunities to behave generously toward others. In a laboratory study involving college students, people were more generous in a continuous Prisoner's Dilemma game when they knew that a third party would be aware of their behavior while choosing between them and others for inclusion in a cooperative task in which they could earn additional money.[85] Outside the lab, possible examples of competitive altruism are widespread, ranging from sponsorship of expensive potlatch ceremonies among Northwest Coast Indians,[86] to charity auctions, to donations of new buildings on college campuses by wealthy philanthropists. A detailed study of public generosity as a hard-to-fake signal was conducted on Mer, an island in the Torres Strait.[87] Although Mer is administered by Australia, the Meriam are culturally and linguistically closer to the people of New Guinea than to Australia's aborigines. Meriam eat a variety of things, including large green turtles (*Chelonia mydas*), each one of which yields about fifty kilos (more than one hundred pounds) of meat. There are two ways to catch turtles. When they are nesting, they can be easily collected off the beaches by just about anybody—men, women, children, and the elderly. Turtle meat obtained that way is mostly shared privately among just a few households. During the non-nesting season, however, the only way to catch a turtle is to head out to sea and capture one in its own element. This is a difficult, risky, and expensive proposition, but some men still do it. Furthermore, when they do it, the meat is never consumed privately but rather widely shared at public ceremonies. Why provide turtle meat—or any other public good—despite the fact that so many free riders will consume it? Because others notice. On average, more than a third of the island's population attends any one feast, which supports the idea that public generosity is a particularly good way to signal one's quality because of its broadcast efficiency. The attention turtle hunters receive eventually results not only in adulation but also in more mates and children.[88]

The example of Mer turtle hunting helps answer a longstanding question about human evolution: How did our ancestors get past the collective action dilemma and become so good at hunting big game? Unlike most other hunting species, human hunters routinely take down prey much larger than themselves. Although it was once thought that this had a direct impact on the hunters' reproductive success by supplying their families with meat, careful ethnographic work among living hunter-gatherers indicates that big game is often much more widely shared. Big game hunting may represent the original public good and, thus, the original collective action dilemma. Kristen Hawkes broke through the haze on this issue with a bold idea called the "show-off hypothesis." Hawkes first pointed out that hunters typically forego small, reliable prey that is shared within families in favor of large, risky game that is more widely shared.[89] She then suggested that the benefit to good hunters was reproductive: Because good hunters are valuable to the whole community and because group membership among hunter-gatherers is usually quite flexible, people have an incentive to make good hunters happy. This may mean taking better care of their children or tolerating the affairs they have with other men's wives. The problem with this is that it creates a second-order collective action dilemma: Why should any one person do whatever it may be that helps keep the good hunter in the community? Why not let someone else carry that burden? In short, why not be a free rider? Recognizing big game hunting as a costly signal breaks us out of this second-order collective action dilemma. If big game hunting is an honest signal of a man's quality as a mate or ally, then receivers who need high-quality mates and allies have their own selfish reasons for attending to it. Although the result of all of this is a public good, only private benefits need be invoked to explain it.[90]

In addition to letting others know about one's quality as a mate or ally, signals can also give them information about one's commitments. The role of commitments in strategic interactions was explored first by Thomas Schelling, who pointed out that it is often in our own long-term best interests to convince others that we will not act in our own short-term best interests.[91] For example, most apartment owners require their renters to sign leases that commit them to at least a year of residence and that specify penalties if the lease is broken. Renters sign leases because they want a commitment from the apartment owner that they will not be evicted if they pay the specified rent. Renters agree to forego opportunities to move to better apartments, and apartment owners agree to forego opportunities to rent apartments to people willing to pay more. Because leases are legal contracts backed by the courts, they represent very believable commitments.[92] A more interesting situation is when there is no legal recourse if a commitment is broken. In that case, signalers must find ways of making their commitments believable to skeptical receivers. Once again, hard-to-fake signals provide a solution.

The question for a signaler is exactly how to make a believable signal of commitment. Receivers sometimes give signalers a helping hand in this regard by specifying what they must do in order for their commitment to be convincing. This is often done by groups that need believable commitments from their members in order to hold together. A common feature of groups is that their members must forego their own short-term best interests either to join or to remain in the group. This can take mild forms such as membership and initiation fees, but many groups require much more onerous signals of members' commitments. Religious groups can be particularly demanding. The kinds of acts and sacrifices required by religion are familiar to everyone: tithing and other contributions of wealth and labor, dietary restrictions, frequent prayer, distinctive clothing, and participation in various rituals, which may involve discomfort if not actual pain.

An intangible but important sign of commitment is often demanded by the very content of religious beliefs. Many religious beliefs involve things that are impossible: virgin birth, resurrection after death, statues that drink milk, lamps that burn for far longer than they should on a given amount of fuel, and so on. To nonbelievers, such things remain impossible. But, precisely because they are impossible, believing in them displays one's willingness to suspend one's rational faculties and thus signals one's commitment. This was clearly recognized by Paul who, in his correspondence with early Christians, encouraged them to embrace beliefs that non-Christians would find "scandalous" and even "moronic."[93] The payoff for religious groups of imposing such costs on their members is that they can be confident that they are truly committed. The payoff to members of paying such high costs is that they get to be members of a cooperative group and to reap the benefits of cooperation.[94]

This idea has recently been tested in two very different religious settings. Richard Sosis and his colleagues compared religious and secular kibbutzim in Israel. Belonging to any sort of kibbutz means fulfilling some group demands, but religious kibbutzim typically expect more signs of commitment than do ones grounded in secular ideologies. Accordingly, religious men made more cooperative decisions in a common-pool resource game than secular men, and religious men who participated in thrice-daily communal prayers were the most cooperative of all.[95] Inspired by Sosis's work, Montserrat Soler conducted a similar study in a very different setting: congregations (*terreiros*) belonging to an Afro-Brazilian religion called Candomblé. Each terreiro is dedicated to one of several gods (*orixás*) in a pantheon derived from African sources. Members of terreiros are required to engage in a wide range of costly behaviors, including sponsorship of large feasts, participation in exhausting dances and other rituals, adherence to a variety of taboos and regulations regarding food and clothing, and completion of difficult and time-consuming initiation rites. Terreiros help members who are in financial trouble, and some even live at the terreiro. Like Sosis, Soler found a correspondence between religiosity and cooperativeness. Terreiro members who made more displays of commitment to the religion made more cooperative choices in a Public Goods Game, and members who were more likely to be in need of support from the terreiro made more displays of commitment. A possible criticism of this kind of work is that it does not distinguish between cooperation that occurs because of all of the signaling that goes on among members of religious congregations and cooperation that occurs because religious believers are afraid of supernatural punishments, whether in this life or the next, if they fail to cooperate. One of the advantages of Soler's study is that in Candomblé there are no beliefs in supernatural punishments or in a judgmental afterlife. Signals appear to be doing all of the work.[96]

The irrationality of religious beliefs and practices enhances their effectiveness as signals of commitment. Because committing oneself to acting against one's own short-term interests is in itself irrational, irrational acts can be convincing displays of such commitments. Robert Frank has suggested that emotions may play an important role in such displays.[97] Consider, for example, romantic love. Committing oneself to a single man or woman is, in most circumstances, irrational. Unless the person in question is of the highest possible quality in every way and will remain that way forever, it is quite likely that someone more desirable, at least in one way or another, will eventually come along. But unless one makes a commitment to stay with a romantic partner even in the face of temptation, one is likely to be quite lonely. Although the commitment requires one to forego opportunities for gain should they arise, it guarantees access to something that is both real and good enough to justify such a trade-off. One way we accomplish this is through the legalities of formal marriage, but it is also accomplished simply through the emotion of romantic love. By blinding us both to our partner's shortcomings and to the competing charms of other prospective mates, romantic love makes our commitment believable. The power of emotional signals is also evident when we compare religious signals, which typically have a powerful emotional element, and secular ones, which typically appeal more to the head than to the heart. This is shown not only by Richard Sosis's findings regarding religious versus secular kibbutzim but also by a comparative study he conducted of nineteenth-century American communes. Religious communes tended both to last much longer and to make more demands upon their members than did secular communes. Furthermore, increasing the number of costly requirements for membership had no effect on secular commune longevity, but it did increase religious commune longevity.

From individuals to organizations

Members of terreiros, kibbutzim, and other religious organizations are able to send signals of commitment to one another because they share not only a common belief system but also a common organizational structure. Such structures include not only religious congregations but also a wide variety of other kinds of groups—firms, bureaucracies, sports teams, and so on. On an evolutionary time scale, such groups are short-lived, which might tempt some evolutionary minded scholars into concluding that we don't need to study them in order to understand cooperation. That might save us all a lot of work (and shorten this book considerably!), but it would be a mistake. Although the individual groups may not last long, the fact that we cooperate within organizational structures is an important fact of human existence. Thus, if we wish to understand cooperation not only on an evolutionary scale but also on a human scale, then we need to understand the important role that institutions and organizations play in making cooperation possible as we go about our everyday lives. That is the focus of the next chapter.

CHAPTER 5

Cooperation and Organizations

SHORTLY AFTER OLSON WROTE *The Logic of Collective Action*, ecologist Garrett Hardin published an essay titled "The Tragedy of the Commons." Like Olson, Hardin was pessimistic regarding our ability to overcome the collective action dilemma. Hardin was inspired by observations made by nineteenth-century economist William Forster Lloyd regarding what happens when people try to share resources. Imagine, for example, herders sharing a common pasture. Each one has an incentive to add additional animals to the pasture because he receives all the benefit while the cost is shared among all. According to Hardin, this set of incentives will inevitably lead to tragedy:

> Each man is locked into a system that compels him to increase his herd without limit—in a world that is limited. Ruin is the destination toward which all men rush, each pursuing his own best interest in a society that believes in the freedom of the commons. Freedom in a commons brings ruin to all.[1]

A conclusion one might draw from evolutionary theory is that, because selection most likely designed us to strive to maximize our own individual fitness, not that of the groups to which we belong, an inability to cooperate is part of our very nature. As we have seen, such a conclusion would be premature at best and simpleminded at worst. Although many animals shape their own physical environments, creating nests, burrows, dams, webs, and so on, humans stand out for the varied and innovative ways in which they have created their own social worlds. These structures include everything from informal rules and cultural norms to organizations, laws, and the state. These human creations, which we will refer to broadly as "institutions," foster cooperation by providing incentives and common frameworks for social action. Whether institutions are as planned and structured as a government bureaucracy or as spontaneous and fluid as a social norm, they matter. Studying human behavior without considering institutions would be like studying animal behavior without considering the environment.

In addition, a central goal of social science is to examine and explain the differences *between* groups of people and across societies. This is one of the reasons why the study of institutions is so important. Humans across the world are genetically very similar, owing perhaps to the fact that we share a common ancestor relatively recently (as compared to say, chimps, a species with a longer history). What genetic variations that do exist occur primarily *within* human populations, not *between* them.[2] We thus are highly unlikely to be able to explain variations that we see when we compare two societies based on genetic variations. Individuals may vary in terms of their innate cooperativeness,[3] but this variation is unlikely to be distributed based on group or region. To explain variation across groups, we must instead look to variations in the environments in which humans live, and these environments are made up largely of manmade institutions.

Institutions, organizations, and norms

While in everyday language the word "institution" means either "a large organization" or "a well-accepted social tradition," in academic parlance it has taken on a more variable set of meanings. Indeed, there seem to be almost as many definitions of institution as there are scholars who study them.[4] Even scholars from the same discipline may define the term in very different ways. To one political scientist, institutions are "stable, valued, recurring patterns of behavior."[5] To another, institutions are "conventions about both language and values" that are "condensed into . . . rules about behavior, especially about making decisions."[6] A third defines institutions as "rules or repeated patterns of behavior that survive the particular individuals who operate them at any one time."[7] Finally, here's another suggestion from the same discipline: "enduring regularities of human action in situations structured by rules, norms, and shared strategies, as well as by the physical world."[8]

With the incisiveness that is a hallmark of his discipline, philosopher John R. Searle has cut through this confusion by using the term to refer to anything that assigns people and objects to statuses that allow them to do things that they would not be able to do solely by virtue of their physical properties.[9] Money, for example, is an institution that is enormously helpful to many types of cooperation, but rectangular pieces of paper and round bits of metal are not.[10] As physical entities, we are simply two organisms. But because we have faculty positions in a university, we have the power to assign grades, approve theses, and so on. Searle's definition is both broad and insightful, identifying the essence of an institution and thus what has led people to use the term to refer to everything from diffuse, decentralized norms and conventions, such as property and marriage, to formal organizations, such as governments and corporations.

As much as we find to admire about Searle's definition, it is often very useful to be able to distinguish between these two broad types of "institution"—informal, decentralized norms and conventions on the one hand and formal organizations on the other. The economist Douglass North has distinguished between formal groups of people (companies, interest groups, governments, and so on), which he calls "organizations," and institutions that are not organizational in form. He uses the word "institution" only for this latter aspect: "Institutions are the rules of the game in a society, or, more formally, are the humanly devised constraints that shape human interaction" and structure incentives in all human activity.[11] In our view, both institutions and organizations are important for understanding cooperation. However, the differences between them are significant enough that we find it easiest to treat them separately. This chapter will focus primarily on the organizational type of institution and the rules that they make, while chapters 6 and 7 will explore the more informal norms and conventions that shape cooperation.

Why organizations succeed . . .

We saw in chapter 3 that institutions in the form of interest groups have helped many groups of people overcome the collective action dilemma. By providing a structure that provides private goods (in the form of social or material selective benefits) to members, groups form that then are able to provide collective goods to all as a by-product of their creation. Although previous generations of scholars pegged the formation of groups to group-level benefits, today most social and evolutionary scientists would agree that such explanations simply beg the question. With biological group selection out of the picture, explanations regarding how and why people form and join groups must refer to their individual evolved and learned goals. To be successful, however, organizations must often deliberately frustrate individuals' efforts to accomplish some of their goals while helping them accomplish others. Organizations that fail to successfully control some of our natural tendencies are doomed to fall apart. For example, our desire for sex is surely evolved, but organizations that do nothing to rein it in are vulnerable to abuses of power, feelings of jealousy, and the trading of sexual favors. Kin favoritism is another good example. A large body of evolutionary theory and empirical research holds that people have a tendency to favor kin over nonkin and close kin over distant kin. In chapter 2, we explained how some organizations take advantage of this tendency through the use of fictive kin terms. Monks are "brothers," American civil rights leaders of the 1960s called upon their "brothers and sisters," and the word "patriotism" itself is rooted in

the Greek for "father." If kin favoritism were the only thing that motivated human behavior, then presumably all human organizations would
be structured around kinship, whether real or fictive. Of course, this is
not the case at all. In fact, many organizations succeed precisely because
they find ways of preventing their members from engaging in kin favoritism, or, more derisively, nepotism. Francis Fukuyama has argued that
while early states were structured around kinship and descent, later em-
pires and, now, democratic governments take pains to make nepotism
difficult to accomplish. The success of the Chinese empire depended
greatly on its system of examinations rather than kinship as a way to fill
slots in the bureaucracy, and modern governments, publicly owned corporations, and other large organizations virtually all have strict rules
against nepotism.[12]

Another reason why some organizations succeed is quite simple: they
build on the successes of previous organizations. For example, political
scientist Robert Putnam and his collaborators have documented how
preexisting civic institutions enabled local and regional democratic institutions in Italy to function more effectively when reforms in the 1970s
gave them new and unprecedented powers.[13] To assess how well the democratic institutions performed, they looked at a variety of indicators:
Does the regional government adopt its budget on schedule? Answer its
mail promptly? Provide day care centers and family clinics? Spend ap-
propriations as planned? And so on. To measure the strength of each region's preexisting institutions, they looked at voter turnout, newspaper
readership, the number of sports and other cultural associations, and an
unusual feature of the Italian electoral system called "preference voting."
Voters in national elections must vote along party lines, but they also
have the option of indicating a preference for a particular candidate from
a party's list. Most voters don't bother to do so, but in some regions competition over preference votes occurs between competing factions within
individual parties. Thus, the frequency of preference voting is an indicator of "personalism, factionalism, and patron-client politics"—in other
words, weak civic institutions.[14] In a statistical analysis of twenty regions,
Putnam and his collaborators found a very strong correlation between
their measures of institutional performance and civic community strength.
The region of Emilia-Romagna, which includes the city of Bologna, does
the best, scoring at the top on both measures. Emilia-Romagna is blessed
with "an unusual concentration of overlapping networks of social solidarity, peopled by citizens with an unusually well-developed public
spirit—a web of civic communities."[15] The preference voting rate in
Emilia-Romagna? Only about 17 percent. Why are some Italians regions
so fortunate? The answer seems to lie deep in Italian history. Emilia-
Romagna and most of the other regions with successful democratic insti-

tutions are in northern Italy, which has a history of rule by communal republics. Those republics, Putnam et al. argue, gave northern Italians a legacy of civic engagement that subsequent generations could build upon.

Because organizations have—by definition—already solved the problem of organizing, and because as a result they have members, rules, and some way of enforcing those rules, existing organizations are often the tools that people turn to when trying to solve some new problem. This may seem obvious in the case of the people of Emilia-Romagna building on their existing state structures or civil rights protestors turning to the churches, but even organizations that themselves are not positive influences on the community might be used by people who want to improve the situation. Sociologist Sudhir Venkatesh, who has spent more than two decades studying the people who live in and near the infamous Robert Wood Taylor Homes housing project on Chicago's South Side, described the way community leaders made use of a local gang to help make their local park safer for their children to play.[16] Street hustlers, prostitutes, and drug dealers held sway in the park, most of them working for (or at least operating with the acquiescence of) the local gang, the pseudonymous Black Kings. The school year was drawing to a close, and parents in the neighborhood were talking among themselves, worrying about where their children would play and how to deal with the expanding illegal activity on their doorsteps. Turning to the police did not seem to be a solution. Police were often unresponsive, and even if they cleared gang activity out of the park once, they could not form the constant presence needed to keep the park safe. Plus, calling the police in had possible negative side effects. Some of the parents themselves had unlicensed business activities in or near the park that might be affected, and there might be backlash from angry gang members. The gang, on the other hand, had the ability to monitor the neighborhood round the clock and the ability to quickly punish any infractions of its rules.

It was decided that the local block club president should meet with Black Cat, the leader of the gang, to try to negotiate a solution. Any confrontation with Black Cat was potentially dangerous, so a well-known local minister brokered the meeting. Before the meeting occurred, residents debated what approach was likely to be most productive. They knew the activities in the park were important to the gang's cash flow, but they also knew that the gang leader needed at least the tacit support of the community if those activities were to continue unimpeded. Perhaps most importantly, Black Cat had aspirations of power beyond the gang, and sought a reputation for being a leader in the community in general. The ideal of "the safety of the children" might help to motivate him. After all, who could possibly argue that the safety of children is not important? The meeting took place, and Black Cat agreed to keep the gang out of the park during daylight hours during the summer. In exchange, members of the block club would convince local women to stop harassing prostitutes, chasing away their clients, and calling the police about drug deals. A local merchant would be convinced to allow drug dealers to congregate in the parking lot near his store instead. The agreement held throughout the summer and into the fall, until the death of Black Cat by a rival's bullet. Why was such cooperation possible? Existing organizational structures played an enormous role. In the beginning, the conversations among the neighbors were facilitated by the fact that there was already a neighborhood block club in place. The leader of that club served as an entrepreneur in the subsequent efforts to come to an agreement with her neighbors and club members about what should be done and then to set up the meetings with Black Cat. The church was another important existing organization, and one that carried with it norms and symbols related to impartiality and peacefulness, helping to provide a modicum of security for the block club leader. But most importantly the existing structure of the gang—created in order to make money, protect turf, and reinforce Black Cat's power—made it possible to provide the public good of a safer park at relatively little cost to the gang. If it were necessary to build such a structure solely to provide a public good, it is unlikely that the community could have provided such a security force. But with the gang structure already in place, and with the concessions from the neighborhood, it cost Black Cat very little to enforce a no-gang policy in the park during the non–school hours of the day. He received the added benefit of solidifying his reputation as an important person in all aspects of the community, and perhaps even feeling that he had done a good deed for "the children."

These examples show how existing social and cultural structures can shape the organizations that foster cooperation. Scholars focusing on social movements have identified three particular such structures as crucial to the success of such movements: political opportunity structures, mobilization structures, and cultural frames. The political opportunity structure is important because social movements have an easier time forming and attaining their goals in some political environments than others. According to sociologist Sidney Tarrow, there are four important dimensions of this political environment to consider: the degree of openness in the formal political institutions, the degree of stability in existing political alignments, the presence or absence of allies, especially influential allies within the governing system, and the presence or absence of conflicts among existing elites.[17] Mobilizational structures are formal or informal means of bringing people together in collective action.[18] Thus they include interest groups, particular methods of protest, and ways of attracting and mobilizing followers. Finally, cultural frames can make it easier or more difficult to mobilize because deeply felt ideas about how society should function or about how individuals should act can influence how people behave. These frames can include existing symbols, cultural understandings and discourses, and any shared meanings within society.[19] During the Solidarity movement in Poland, for example, symbols stemming from the populace's shared Catholicism helped galvanize opposition and challenge the legitimacy of the government.[20]

A close look at the U.S. civil rights movement can help make these ideas a bit more clear. The stage was set by the economic and political changes in the United States that preceded the movement's beginnings in the mid 1950s: African American migration from the rural South to cities and towns, both in the South but especially in the North, the growth of the African American middle class, and the growing affluence of post–World War II America. These led to political opportunities in the form of political parties that hoped to capture this potential new voting bloc. Thanks both to federalism and the separation of federal power into legislative, executive, and judicial branches, the United States has always been a relatively porous system with multiple venues for potential access. The civil rights movement made use of this porosity, protesting at the local level, filing lawsuits at multiple levels, and, after a decade of effort, pushing for legislation at the national level, which culminated in the Civil Rights Act of 1964 and the Voting Rights Act of 1965. African American churches were a key mobilizing structure. For example, during the 1954 bus boycott in Montgomery, Alabama, meetings of activists took place in the First Baptist Church. Just a few blocks away was the Dexter Avenue Baptist Church, where the young Martin Luther King, Jr. was a minister. Although the movement might have formed even without the help of churches, they would have had to find other places to meet, others willing and able to serve as leaders, and new ways to quickly reach out to large numbers of people within the community. The churches—not only in Montgomery, but throughout the South—provided the places, the leaders, and the communication networks. Not only formal organizations, but means of organizing can be considered mobilizational structures, and thus in the case of the civil rights movements we would include the repertoire of pacifist protest techniques that were taught to activists in the movement that allowed them to remain calm and non-retaliatory when police and bigoted mobs would attack their sit-ins and boycotts. There were many useful frames and symbols within the movement, but perhaps most important is the "rights" frame, that is, the idea that all that they were asking for were basic human rights that were guaranteed to them in the Constitution. The rights frame became all the more convincing after the Supreme Court's *Brown v. Board of Education* decision in 1954 in which school segregation along racial lines was declared unconstitutional. The *Brown v. Board of Education* decision is sometimes belittled by political scientists who point out that it did not force any change.[21] In fact, desegregation did not begin until after Congress stepped in with the Civil Rights Acts, ten years after the court ruling. However, the decision provided a legitimacy to the movement that otherwise would have been lacking. The Supreme Court had essentially said that the movement was in the right and anyone opposing them was unconstitutional. Thus, the rights frame was integral to the success of the movement.[22]

Of course, environments are not static. Because humans create their own environments, they can affect their opportunities and the ways issues are framed through their own actions. Protests can lead elites to begin paying attention to an issue or cause schisms among elites. These become new political opportunities. Although it is always easiest to use existing mobilizational structures, new interest groups and social movement organizations can be built from the ground up. Frames can be shifted as well. The "consciousness raising" of the 1960s was an attempt to reframe the way people thought about existing power structures so that they were more likely to act against them. Thus, cultural frames are not everlasting or unchanging but a dynamic part of the process. Viewed at any one point in time, these elements may appear fixed, and indeed they may endure for years, decades, and even centuries. But the possibility is always present that action will produce change in these structures.

The voluminous literature on common-pool resources provides additional insights into why some organizations are successful in fostering cooperation. Common-pool resource management is informative because such resources are so vexing. To understand why, compare them to public goods. Public goods pose a problem because it is hard to exclude free riders from enjoying them. Public goods' saving grace is that even though free riders are difficult to exclude, at least those free riders do not prevent those who contribute to the public good from enjoying it, as well. National defense, for example, is a good consumed not only by those who pay taxes but also by those who do not, but those who do not pay also do not reduce its effectiveness for those who do. In the jargon of economics, public goods are nonrivalrous. This contrasts with common-pool resources. In addition to pasture, examples of common-pool resources include commonly owned irrigation water, forests, and fisheries. As with public goods, it is hard to exclude people from consuming common-pool resources even if they have not contributed to their production or maintenance. What makes common-pool resources doubly vexing is that, like private goods, they are also rivalrous, that is, one person's consumption of the good reduces the amount that is available for others to consume.

Because common-pool resources are both problematic and widespread, they have been the subject of a great deal of theoretical and empirical research. Until fairly recently, scholars tended to advocate one of two extreme approaches to the problem of common-pool resources: government and the market. The government can solve the problem of how to manage a common-pool resource by creating a set of rules and imposing them from the outside. Faced with depleting fish stocks, for example, the government may step in and create rules about how much each fishing boat may catch during particular periods of time. At the other extreme, some common-pool resources can simply be divided up and turned into private goods. Shared pasture, for example, can be parceled out to individual livestock owners rather than shared by a community.

Many communities, however, have found a middle way to manage common-pool resources, one that avoids both external governmental controls and privatization. These common-pool resource management schemes have been well documented by anthropologists, political scientists, and economists, and they tend to share certain key features. Much of this body of work has been synthesized and summarized by political scientist Elinor Ostrom and her colleagues. Ostrom has pointed out that efforts to manage common-pool resources face three basic problems. First, there is the problem of how to supply the new rules and mechanisms for enforcing them that will allow those who share the resource to use it but not abuse it. Creating such rules is costly and time-consuming, as is enforcing them. Furthermore, not all rules will have similar impacts on all participants in the management scheme, so they may disagree about the details. Second, there is the problem of how the participants in the management scheme can make their commitments to that scheme credible to each other. Given that it may sometimes be very tempting to break the rules (e.g., saving one's crops during a drought by drawing off more irrigation water than one is allowed), such commitment is crucial to the success of any common-pool resource management scheme. In the absence of credible commitment, participants will be skeptical regarding other participants' willingness to adhere to the rules, and the agreement may fall apart. Third, there is the problem of mutual monitoring: Participants in a common-pool resource management scheme need to be able to see whether others involved in the scheme are following its ground rules. Mutual monitoring is related to the problem of credible commitment. If it is easy for participants in the scheme to monitor one another's use of the resource, then credible commitment is less of a problem. But if mutual monitoring is difficult, meaning that cheating is easy, credible commitment becomes more important.

Despite these and other problems faced by those who would collectively manage common-pool resources, many good examples do exist of communities with long-lasting schemes for doing just that. These range from irrigation schemes in Spain and the Philippines to community-managed grazing lands in places as diverse as Switzerland and Japan. One particularly detailed portrait of such as scheme was provided by ethnographer Robert McC. Netting in his book *Balancing on an Alp*.[23] Netting studied Törbel, a village in the southern Swiss canton of Valais (or, to the German-speaking people living in Törbel, Wallis). Törbel is an ancient community, first mentioned in writing in the eleventh century. Törbel's farming economy is centered on nuclear family households that grow rye, fruits, and vegetables and keep livestock. Grain fields, hay fields, gardens, and livestock are all privately owned. However, following an agreement first put in writing in 1483, Törbel's citizens communally own and manage its paths and roads, its irrigation system, its forests, and its summer pastures in the high alpine meadows above the village. Summer pastures were successfully managed thanks to a strict rule that each citizen could send up the mountain only as many cattle as he could feed through the winter with the hay he had grown during the summer. Because the use of the summer pastures was tied to the availability of winter forage, overgrazing was avoided. Other written rules governed other aspects of Törbel's common property management scheme. For example, the men who tended the livestock in the summer pastures were responsible for making cheese, which was then distributed to each household in proportion to how many cows it had sent to the pasture. The community also made arrangements for the summer pasture to be fertilized with manure and controlled access to firewood and timber from the community-owned forest.

One thing that helped make the Törbel scheme a success was that it was closed. Simply buying land in the village did not give one rights to use the village's common-pool resources. Thus, from the point of view of an outsider, Törbel's common-pool resources look a lot like private property. This is a common feature of successfully managed common-pool resources. Consider, for instance, privately owned urban parks, such as New York City's Gramercy Park. Only people living in buildings that face the park can buy keys to its gates. From the point of view of everyone else, then, Gramercy Park is privately owned. But from the point of view of those who live along the park, it is a common-pool resource. Despite the importance of communally held property to Törbel's economy, Törbel was not a commune. Private property rights were well developed and respected. Why are some kinds of property held privately and others communally? According to Ostrom, communally owned lands tend to have low rates of production and undependable yields. This means that a large area is needed for them to be used effectively, which makes them hard for a single owner to defend. Grazing land often fits this profile but cropland usually does not, which may explain why communally managed grazing land is so much more common than communally managed cropland.

. . . and why they fail

Cooperation's success stories are better studied and perhaps more enjoy-able to read about than its failures, but it is only by contrasting the suc-cesses with the failures that we can identify the characteristics that make organizations work. Consider, for example, Putnam et al.'s work on dem-ocratic institutions and civic engagement in Italy. The success story of Emilia-Romagna seems all the more remarkable when contrasted with other Italian regions in which democratic institutions perform poorly and civic engagement is weak. While Emilia-Romagna was graced with a his-tory of communal republics, Calabria, the toe in the boot of the Italian peninsula, was ruled for centuries by kings, many of them foreigners. James Walston, an expert on the region, has noted that "[t]rust is not a commodity in great supply" there and that "the lack of trust affects all types of social interaction except those based on links which can be ex-ploited to mutual advantage."[24] Not surprisingly, the rate of preference voting in Calabria is about three times that in Emilia-Romagna.

Calabria's lack of trust and civic associations that foster it is not un-common. The same problems may help explain another failed organiza-tion, this one on the Honduran island of Utila. In 1996, Lee and his stu-dent Shannon Steadman were conducting fieldwork on Utila, one of Honduras's predominantly English-speaking Bay Islands. Although they were on the island to study how relationships are negotiated within Util-ian families,[25] they were fortunate to be there immediately after some of the islanders had attempted to collectively manage a common-pool resource.[26]

Most of Utila's population of just over two thousand can be traced back to English-speaking settlers of European and African descent who started arriving on the island in about the 1830s, chiefly from the Cay-man Islands and other places in the Anglophone Caribbean. For most of the twentieth century the Utilian economy was dominated by remittances sent back by men working on ships, mainly in the tropical fruit and oil industries. But starting in the 1960s, a small tourist industry began to grow on the island. Because Utila's beaches are small and unimpressive, the tourist industry focused instead on the many beautiful reefs that sur-round the island. By the time we arrived on the island, there were twelve dive shops operating, divided roughly equally between those owned by foreigners and those owned by Utilians. All were small and focused pri-marily on offering inexpensive scuba diving lessons to the many back-packing tourists who came to the island in the winter and summer, when college students from the United States and Europe were out of classes. It was not uncommon to find shops offering the most basic class for as little as $100. Some shops were affiliated with lodges and restaurants and so could offer package deals. Another common way to compete was to offer one or two free dives after a diver passed the course and received PADI (Professional Association of Diving Instructors) certification.

The reef itself is obviously an important common-pool resource shared by Utila's dive shops. Because one of the easiest ways to damage a reef is to drop an anchor on it, Utila's dive boats instead tie off at one of dozens of mooring buoys around the island. The buoys were placed originally by the Bay Islands Conservation Association, a nonprofit organization funded by international conservation organizations and other sources. The buoys are now maintained by the municipality of Utila, which charges divers a daily fee. This is an example of how an external authority can manage a common-pool resource on behalf of those who use it.

A less obvious common-pool resource shared by Utila's dive shops is the island's tourists. Although we do not often think of the customers of a particular kind of business as a common-pool resource, there is no reason not to do so. Such customers do have the two requisite characteristics of all common-pool resources, low excludability (it is hard to prevent either new firms from entering the market or existing firms from lowering their prices) and high subtractability (money spent by a customer at one business is not available to other businesses). A possible objection is that collusion among firms to fix prices is not a "good," but that is simply a confusion of two meanings of "good." There is nothing about an economic "good" that implies that it is a moral "good." Certainly, keeping prices higher than they would be under conditions of full competition among Utila's dive shops would be good—economically—for the dive shops themselves so long as they did not damage the island's ability to compete with competing dive sites, such as the neighboring island of Roatán.

Utila's dive shop owners recognized that their profits were being damaged by cutthroat competition, and in June 1995 they verbally agreed to fix the cost of a basic scuba diving course at $139. An early sign that this agreement was doomed was that one dive shop owner immediately slashed his price to $99 in a fit of pique at the idea that the other owners would attempt to tell him how to run his business. Because June was the beginning of one of Utila's two high seasons and tourists were plentiful, the other owners stuck to the agreement. The situation started to change in September, when American and European college students returned to classes and the dive shops were having a hard time finding customers. Shops first started competing without violating the pricing agreement by offering extra free dives with classes, but they soon began to cut prices, as well.

In November 1995, the owners made another attempt to fix prices, this time in writing. They formally established the Association of Utila

Dive Shop Owners (AUDSO) and agreed to fix the price of a basic scuba diving course at $149. Prices for individual dives and packages of dives were also set, and the shops agreed to give out no free dives with courses. One practice that irked some of the owners was the solicitation of customers at the island's airstrip and dock. They argued that, because Utila's tourists typically knew little about either the island or diving when they first arrived, urging them to commit to one dive shop before they had had an opportunity to do some comparison shopping was taking advantage of their ignorance. The owners agreed not to solicit customers at Utila's dock and airstrip but encouraged everyone to solicit customers on the Honduran mainland in the hope that this would steer more customers to Utila—and all its businesses—and away from Roatán and other competitors. In order to keep all of the dive shops small, they agreed never to hire more than five dive instructors. The AUDSO was also charged with establishing a safety committee to monitor shops' compliance with basic safety standards. Member shops were required to pay a fee of $30 per month. To further reduce competition, the AUDSO asked Utila's mayor, who was also part owner of one of the shops, to refuse permits to any new dive shops. He agreed to do so and also noted the relevance to AUDSO's efforts to fix prices of an obscure Honduran law requiring businesses to post price increases one month prior to the actual increase.

The AUDSO had its final meeting in January 1996. By the time that Lee and Shannon arrived on the island in June of that year, the organization and its price-fixing agreement had all fallen apart. Through interviews with the dive shop owners, we were able to reconstruct what had happened. One of the AUDSO's greatest shortcomings was its failure to establish an official way of monitoring member shops' compliance with the agreement. Shops therefore resorted to informally policing each other, which exacerbated existing tensions among shops and led to new ones. The mayor also reneged on his agreement not to issue licenses for new dive shops, arguing that he was not actually in a strong legal position to do so. And there was also the ambiguous legal status of such a price-fixing agreement. Although the cultural distance between the English-speaking Bay Islands and the Spanish-speaking Honduran mainland means that Honduran law has not always been strictly enforced on Utila, any agreement to fix prices might eventually have come under the scrutiny of Honduran authorities.

Why has Törbel's common-pool resource management scheme lasted for half a millennium while Utila's fell apart after a few months? Based on comparisons between common-pool resource management schemes around the world, Ostrom has identified seven design features that are common to those that last. First, they must have clearly defined boundaries. While the membership in Törbel's agreement was clearly limited by its charter, new dive shops opened up on Utila quickly and easily despite the AUDSO's efforts to prevent them from doing so. Second, the rules of the agreement must be appropriate to local conditions. Törbel's stipulation that a cattle owner could send only as many animals to the summer pasture as he could feed during the winter used local conditions to limit exploitation of the common-pool resource. This is one instance in which the AUDSO got it right. They recognized two important characteristics of their customers—they were on tight budgets and did not know very much about scuba diving. The AUDSO agreement recognized these characteristics by keeping prices low compared to most other places in the world and by discouraging shops from taking advantage of naïve customers by soliciting business at the dock and airstrip before the arriving tourists had had a chance to become familiar with the island and do some comparison shopping. Third, those who are affected by the rules must be able to participate in changing them. While Törbel held annual meetings to manage its resources, the AUDSO set up neither a schedule of meetings nor any procedure for modifying its agreement.[27] Fourth, some provisions must be made for monitoring members' compliance with the agreement. In Törbel, it was easy to make sure that no one sent too many cattle to the pasture simply by counting them as they left the village. On Utila, monitoring was left up to the individual shops. Fifth, there must be graduated sanctions for violations of the rules. Törbel established a clear system of fines for those who violated the rules by, for instance, cutting trees in the forest without permission. The AUDSO established no system of sanctions other than a vague notion that they would remove the AUDSO certificate from the window of any shop that failed to live up to the agreement. Sixth, there must be a system for conflict resolution. Törbel's system of annual meetings allowed disputes to be settled in a peaceful way. The AUDSO established no system for conflict resolution. Finally, the seventh rule is that there must be no external limits on a community's efforts to manage a common-pool resource. While the AUDSO was pushing the envelope regarding what Honduran law would allow, Törbel's isolation meant that external interference in its affairs was never a worry.

The failure of the AUDSO also points to the importance of cultural context in shaping common-pool resource management schemes and other organizations that try to foster cooperation.[28] Utilians are perhaps the most individualistic people in the world. They themselves acknowledge that this often makes it difficult for them to organize not only things like AUDSO but also more mundane activities like sports teams and leagues. The fact that half of the dive shop owners are non-Utilians does not help because many of them came to the island in an attempt to get away from the kinds of rules and regulations that irritated them in their home countries. Such extreme individualism is so common in the ethnographic literature on Caribbean societies that it even has a name: crab antics.[29] The analogy is to a barrel full of crabs. If the crabs were to work together, they could all escape. But there is no need to put a lid on the barrel because the crabs will never work together. Instead, as one crab reaches the mouth of the barrel, another will pull it down in an attempt to pull itself up.

In a sense, then, Utila is an Olsonian antiparadise, a place filled with precisely the sort of "rational, self-interested individuals" that are so central to his critique of previous theories about group formation and collective action. But of course we aren't all Utilans. Some of us live instead in Törbel and other communities in which individualism is muted or even aggressively suppressed. These existing cultural tendencies are not something that can be changed simply by one person—or even a group of people—wishing it changed. Institutions both formal and informal are built on an existing history of organizations and shared expectations—in the words of Douglass North, they are "path dependent."[30] What has gone before influences what comes later. Residents of Venkatesh's South Side Chicago neighborhood could not simply choose to have a non-gang-related security force protect their park; they had to work with the materials and organizations they had at hand. Institutions, especially those with organizational form, are also "sticky"—once they have been created, they tend to hang around. That, of course, does not mean that it is not possible to change them, but it does mean that change usually requires a sustained period of effort toward that change, as with the civil rights movement in the United States. The more bureaucratized and stickier the institution is, the more pressure is needed for change to occur, and when change does arrive it can do so very suddenly.[31] In the meantime, would-be human cooperators must find a way to do work toward cooperative outcomes within the historical, cultural, and institutional contexts in which they live.

Thinking like a team: Cultural group selection and coalitional psychology

Organizations compete with each other for members, territory, profits, and other resources. This competition sets the stage for a process of variation and selection that is analogous to how selection acts on organisms. Such "cultural group selection" is an important new frontier in the evolutionary study of cooperation. The phrase "cultural group selection" can be parsed in two ways: "{cultural{group selection}}" and "{{cultural group}selection}." The difference is subtle but important. In the first case, cultural group selection appears to be a type of group selection, analogous, perhaps, to biological group selection. In the second, it appears to be selection among groups defined in terms of cultural similarities and differences. We encourage readers to think of it in the second way, that is, as a type of selection that acts on groups that are defined in terms of shared culture rather than as a type of group selection that happens to involve culture in some way. This helps separate cultural group selection from biological group selection. This is useful not only because of biological group selection's mixed intellectual legacy but also because the two theories are, in fact, quite different and have very different implications for the study of human cooperation.

The general idea of cultural group selection has been around for a long time. Versions of it can be found in the writings of nineteenth-century evolutionists and in twentieth-century cultural anthropology's structural-functionalist and neofunctionalist schools of thought.[32] More recently, the idea was proposed by such eminent scholars as Donald T. Campbell and Friedrich A. Hayek.[33] However, it was cultural transmission theorists Robert Boyd and Peter J. Richerson who formalized and developed the theory and their students, most notably Joseph Henrich, who have brought it to center stage in the study of human cooperation.[34]

The basic ideas behind cultural group selection are straightforward. There are patterns in how people acquire culture, such as a tendency to conform to local culture traits and a tendency for common versions of traits to be copied more often than rare ones.[35] These patterns tend to create cultural similarities within groups and differences between them. Some of these cultural differences will lead some groups to outgrow, outlast, and generally out-compete other groups. Some groups may even go extinct. Although this sounds a lot like biological group selection, the circumstances it requires are common rather than rare. While biological group selection is more likely when migration between groups is limited and groups are discreet, migration is not a problem for cultural group selection so long as migrants learn new culture traits when they move to new groups. Indeed, one way for one cultural group to outcompete another is simply to draw off its people. It is clear that this possibility was understood by Soviet leaders when they authorized extreme and often lethal measures to prevent citizens of Warsaw Pact countries from fleeing to the West.

Cultural group selection differs from biological group selection in one other very important way, as well: evidence for it is indisputable. Indeed, it happens every day. Our favorite example of cultural group selection is competition among firms in a market economy.[36] Even if they provide the same product or service, firms differ from one another. Because those differences certainly are not genetic, they must be cultural, that is, the result of socially transmitted information. Some of the cultural differences between firms may be trivial, but some may have real effects on the bottom line and thus on firm longevity. Although the scholarly literature on organizational culture is certainly relevant here, the sorts of cultural differences likely to make a difference when firms compete in a market also include manufacturing methods, labor-management relations, and marketing techniques. Firms work hard to refine these aspects of their internal cultures, and when they find something that works for them, they also work hard to keep it out of the hands (and minds) of people in competing firms. Such vigilance helps maintain the differences between groups that are necessary for cultural group selection to operate. Although this process drives rapid cultural change in market economies, the individuals involved are still very much individuals, free to move from firm to firm, acquiring new culture traits with each new job.

One thing that remains fairly vague in the literature on cultural group selection is what constitutes a "cultural group." Firms certainly qualify, but they have several characteristics that may make them rather unusual. Competition among them is intense, with firms being founded and dying out with great frequency. Except in cases of mergers and acquisitions, firms are normally quite discrete. Firms are functionally integrated and have clear corporate structures. Needless to say, not all groups are like this. Indeed, some "groups" might better be thought of as categories. Consider ethnic "groups," for example. Particularly among the nonstate societies documented by ethnographers, such "groups" are really just categories, that is, people who share a bundle of culture traits, typically including a language. Although they may have local institutions and organizations, as a whole they do not necessarily have any sort of functionally integrated corporate structure. Do such loosely defined "groups" suffice for cultural group selection? Perhaps. This is an area of theory that is greatly in need of clarification. Our sense is that cultural group selection will have an easier time having a noticeable effect on the distribution of culture traits within and between groups if the groups are more like firms (discrete, functionally integrated, etc.) and less like ethnic groups.

The kinds of culture traits that will be selected are also likely to differ depending on the kinds of "groups" in question. Selection among firms or other corporate, functionally integrated groups will largely be on culture traits that affect them as groups, that is, traits that influence their ability to achieve their group-level goals. When cultural group selection occurs among "groups" that are really just categories, no such traits exist. Instead, such groups differ in terms of the extent to which the culture traits that are prevalent within them help their members to survive and reproduce. For example, some ethnic groups may succeed and others fail through cultural group selection because some happen to have culture traits that help their bearers survive and reproduce and that are difficult to disentangle from other traits that go along with the group's ethnic identity. Since the 1980s, Lee has been studying just such a process among Maa speakers in East Africa. The most famous Maa speakers are the Maasai, but a variety of other peoples, scattered mainly up and down East Africa's Great Rift Valley, also speak the language. Although Maa speakers do a variety of things to make a living, including fishing, beekeeping, farming, and foraging, the Maa language and the Maasai themselves are closely associated with pastoralism. The Maasai are very successful pastoralists. Their success undoubtedly stems from a wide range of culture traits including, but not limited to, their knowledge of animal husbandry. Also important is their age-set system, which provided the manpower and organization to defend herds and grazing lands, and a system of risk pooling through gifts of livestock called *osotua* (we will have more to say about osotua in chapter 6). As a result of their success, Maasai identity is an attractive one, and a variety of other groups have aspired to it. We have studied one such group, the Mukogodo of Kenya. In just a few years in the early twentieth century, the Mukogodo radically changed their way of life. This included the loss of their old residential pattern (living in rock shelters), their old subsistence pattern (beekeeping and foraging), and their own language (Yaaku) in favor of the Maasai alternatives (living in houses made of cow dung, raising livestock, and speaking Maa).[37]

In between the extremes of firms and ethnic groups lies a wide variety of groups that may be subject to cultural group selection. Most interesting to us are the shifting, temporary coalitions that form so important a part of our everyday lives in spheres ranging from work and politics to recreation and religion. As we have seen, gene-culture coevolution may have shaped adaptations, such as an ability to identify social norm violators, that promote cooperation.[38] That example involved selection on individual culture traits and on biological individuals rather than on cultural or biological groups. It is also possible for cultural group selection to create selection pressures at the individual level that help shape evolved human psychology. What remains unclear is exactly what kinds of psychological predispositions would be favored by cultural group selection. Joseph Henrich has suggested that the result will be a broad range of "prosocial" predispositions that are the foundations of "large-scale cooperation" among humans.[39] We find "prosocial" to be too general a category to be a useful starting point for the specification of psychological adaptations. "Large-scale cooperation" is a broad category that seems to refer primarily to states and other large organizations, to more diffuse systems of cooperation, such as the market, and to more specific phenomena such as "voting, giving blood, food sharing, not extorting each other, policing, and territorial defense."[40] As others have pointed out, the idea that states and markets depend upon "prosocial preferences" is quaint, at best.[41] Indeed, one could easily argue that states and markets have been so successful precisely because they rely not upon prosocial preferences but rather upon quite selfish ones: in the first case, a desire to avoid being punished and, in the second, a desire for profit.

The kinds of psychological adaptations that we see as most likely to arise from gene-culture coevolution in the context of cultural group selection were captured a few years ago by the comedian Jerry Seinfeld. He pointed out that, because players frequently move from team to team, sports fans are "actually rooting for the clothes . . . you are standing and cheering and yelling for your clothes to beat the clothes from another city."[42]

The world created by cultural group selection is one in which competition between groups is important, but group memberships are flexible rather than fixed. This puts pressure on people to be team players but to recognize that they are just that—teams. People need to be able to feel loyalty and commitment toward their current teams while also being willing and able to shift from team to team. In short, we should have psychological mechanisms designed to deal with a world of important but flexible coalitions. In Clark and Wilson's terms, we should experience "solidary benefits" when we work with others toward common goals. This is similar to a suggestion made by economists Robert Sugden and Michael Bacharach that much human cooperation rests on our ability to "think like a team."[43] While such "team reasoning" may indeed be very important, it is unclear where it comes from.

Cultural group selection combined with gene-culture coevolution may provide an answer. A reasonable hypothesis is that our evolved coalitional psychology is built upon psychological adaptations originally shaped by direct and indirect reciprocity.[44] This idea also emerges from a model of human evolution developed by Richard Alexander, Mark Flinn, and Kathryn Coe, who have suggested that the driving force behind the evolution of human uniqueness was an evolutionary arms race between coalitions combined with an "informational arms race" created by our species' growing dependence upon culture.[45] Although Alexander and his colleagues did not relate their idea to the theory of cultural group selection, it is a short step from one to the other.

Several recent studies provide support for this view. At Lancaster University in the United Kingdom, social psychologist Mark Levine and his coauthors focused on football (soccer) fans, in particular fans of Manchester United.[46] In the first study, subjects who had already been identified as fans of Manchester United were given a series of questionnaires to heighten their sense of identification with the team and with their fellow fans. They were then told that they would have to be taken across campus for the second part of the study, which involved watching a video about football and its fans. As they were walking across campus, a researcher playing the role of a jogger fell down, grabbed his ankle, and shouted as if in pain. The experimental condition was in which of three shirts the jogger was wearing: a Manchester United shirt, a plain shirt, or a shirt branded with the logo of Manchester's bitter rival, Liverpool FC. All but one of the subjects who saw a fellow Manchester fan fall down came to his aid, but they helped the runner in the plain shirt only a third of the time, and they helped the Liverpool fan even less often. In the second study, Levine and his colleagues again recruited Manchester United fans, but this time they gave them questionnaires that primed their sense of being football fans in general rather than Manchester fans in particular. This time, both the Manchester United and Liverpool FC shirts elicited high rates of helping compared to the plain shirt. This is all the more remarkable given how bitter—and sometimes even violent—the rivalry between the two teams has often been. If Jerry Seinfeld were a psychologist, he himself could not have designed a better experiment for demonstrating both the importance and the flexibility of our coalitional identifications.

Shirts once again played an important role in a study by psychologist Robert Kurzban and his colleagues on which cues people use when identifying people as members of coalitions.[47] Because coalitions are flexible, people should be able to pick up cues that are easily changed, such as clothing and jewelry, as well as those that are more fixed, such as accents and physical resemblances. Their subjects were shown photographs of members of two rival basketball teams and told to form impressions of the individuals on the teams. Each picture was paired with a statement that the person had supposedly made about the teams' rivalry. The actual pairing of sentences with photos was randomized across subjects. Subjects were then given a surprise memory test involving matching statements with photos. Because this was a difficult task, they made a lot of errors, and the patterns in the errors reveal that they used statements associated with faces along with other cues, such as the basketball jersey colors, to identify coalitions. One of this study's most interesting findings is that flexible cues such as the statements people make and the clothes they wear swamp the effects of race as a coalitional cue. This makes sense in light of how our ancestors lived. Given that their mobility was limited by how far they could walk, they were very unlikely to have encountered people as physically different from themselves as we routinely do now, and it would make little sense for us to have an evolved tendency to focus on race when determining coalitions. Kurzban et al.'s encouraging conclusion is that racism may simply be a misfiring of a psychological mechanism designed to pick up on more flexible coalitional cues. When race is

disconnected from actual coalitions, as in their experiment, its importance as a way to sort people into groups is greatly diminished. It may be
that sports fans who root for clothes rather than players have it right,
after all: it's the flexible coalitional cues that really matter.

Another laboratory study demonstrates what managers, coaches, and
military officers have long known: if you want to get people to work successfully together in a group, give them a rival group with which to
compete. Mikael Puurtinen and Tapio Mappes had their subjects play a
Public Goods Game under two different conditions.[48] Some played a
simple Public Goods Game. Others played the same game, but with a
twist: after each round, groups were paired, and whichever group had
higher contributions to the common pot received the difference between
the two groups' common pots, multiplied by two. The group with the
smaller common pot lost the same amount of money. Not surprisingly,
contributions in the group competition condition were much higher than
in the control condition. Puurtinen and Mappes also asked their subjects
questions about how they perceived the games and their emotional states
while they played them. Subjects in the control condition saw the other
players as competitors; those in the group competition condition saw
them as collaborators. Players in the group competition condition also
reported more anger toward free riders and more guilt about their own
selfish behavior than players in the control condition. Because subjects
were anonymous and had no ability to punish each other, such emotions
make no rational sense. But they do make evolutionary sense as the kinds
of emotions that motivate cooperative behavior during competition between coalitions, even those as fleeting as the coalitions formed artificially in Puurtinen and Mappes's laboratory.

Two recent field projects also shed light on our evolved coalitional
psychology. In the New Guinea highlands, Helen Bernhard ran Third-
Party Punishment Games with members of two different ethnic groups,
the Wolimbka and the Ngenika.[49] Although the Wolimbka and Ngenika
do not have any history of warfare with each other, they normally have
very little interaction with each other. In Bernhard's experiment, players
were informed about the ethnicities of the other two players. Dictators
expected to be punished less by a member of their own group than by a
member of the other group expected more punishment for low offers
when the recipient and the audience were in the same group than when
they were not, and gave less when the audience was a member of their
own group. Overall, punishment for low offers was greatest when the
audience and the recipient were members of the same group and the dictator was not. Punishment levels were also high when all three were
members of the same group, suggesting a desire to enforce a sharing
norm. Punishment levels were lowest when the dictator was in the same

group as the audience but the recipient was not, suggesting a lenience regarding the sharing norm when the perpetrator is a member of one's own group, and when the audience is not of the same group as the dictator and recipient, suggesting a reticence to spend money to enforce another group's norms. Bernhard et al. used the term "parochialism" to describe these patterns, but their results fit well within the framework of evolved coalitional psychology.

In the Ecuadorian Amazon, anthropologist John Q. Patton has collected detailed ethnographic data on coalitions in a community of horticulturalists and foragers.[50] The community, Conambo, contains speakers of three different languages (Achuar, Quichua, and Zapara) and members of two different political coalitions. Because violent conflict is common, with half of all men appearing in the genealogies of Conambo's families having died in homicides, being a member of a strong coalition is an important social asset. Coalitional identities help shape individuals' social lives in important ways, and individuals, in turn, contribute to the strength of coalitions. For example, when asked for their view of others' status in the community, residents of Conambo bias their responses in favor of people in their own coalitions. Residents of Conambo also share meat, which is a large part of their surprisingly rich diet, more often with coalition partners than with members of the community's other coalition. Furthermore, men in Conambo who share more meat with each other are thought by other members of the community as more likely to support each other when conflict occurs. Finally, in an experiment using the Ultimatum Game, Patton found that Conambo residents with stronger coalitional ties in the community tended to make more generous offers.

Although it might be tempting to call Conambo's coalitions "groups," they are not rigid enough to warrant that term. The two coalitions in Conambo do roughly correspond with Achuar speakers on one hand and Quichua speakers on the other, but there are many mixed households and individuals who switch from one ethnic identity to another depending on the audience present. Although "tribes" are often invoked in the theoretical literature on cultural group selection, we may be better off focusing instead on more temporary and ephemeral coalitions. In fact, ethnicity itself, though often imbued with species-like qualities,[51] can be quite malleable and fluid. For example, the Mukogodo people we have been studying since the 1980s are particularly adept at shifting their ethnic identity according to the audience present. Although they usually call themselves Mukogodo, they also refer to themselves as Dorobo, Maasai, and Yaaku, depending on whether they are trying to influence British colonial officials, Maasai speakers of higher status than themselves, or, most recently, international indigenous rights activists.[52] Ethnicities are also surprisingly ephemeral. Where, for example, have all the Picts and Visigoths gone? Although they certainly left behind biological descendants, no one uses those labels to refer to themselves any more. On the other hand, some currently important and seemingly ancient ethnicities are actually quite recent inventions. Only since the nineteenth century has "French" been a meaningful ethnic label for more than a small minority of people in the nation of France.[53] The flexibility and ephemerality of ethnic identity suggests that our evolved psychology for dealing with group life is likely one designed to deal with coalitions that may come and go, rather than long-lasting, bounded, and distinct groups.

From groups to norms

It is clear from the discussion of cultural group selection above that sometimes even the *suggestion* of an organization is enough to get people to cooperate. This observation leads us to the topic of those other kinds of "institutions," the ones that are not formally organized but that still help shape the ways humans cooperate by giving us common frameworks in which to do so. Although these still fit some definitions of "institution," most scholars refer to them instead as "norms" or "conventions." Because norms and conventions are crucial to our understanding of the phenomenon of human cooperation, we have devoted two chapters to them. The next chapter focuses on the ways in which common knowledge helps people to coordinate their actions, even with strangers.

CHAPTER 6

Meeting at Penn Station

COORDINATION PROBLEMS AND COOPERATION

YOU AND A FRIEND need to meet in New York City, but you did not prearrange a time or place. Where would you go, and at what time of day? Yale economist Thomas Schelling presented that question to "an unscientific sample of respondents" in New Haven, Connecticut, while writing his seminal book *The Strategy of Conflict*. A glance at the title of our book will give you part of the answer: A majority chose the information booth at Grand Central Terminal. And virtually everyone chose noon. Grand Central and twelve noon were prominent, salient focal point solutions to the problem of coordinating a meeting time and place.

As we have seen, two kinds of problems can stymie cooperation: collective action dilemmas and coordination problems. Schelling's meeting place scenario is an example of a coordination problem. While conflicts of interest make cooperation difficult in collective action problems, the main roadblock to cooperation in coordination problems is simply a lack of common information: everyone would benefit if coordination were to be achieved, but they may not know how to do it. Although most research on cooperation to date has concerned collective action, a growing number of scholars—including us—think that coordination problems deserve more attention.[1]

Edna Ullmann-Margalit illustrated the difference between collective action and coordination problems with an imaginary game involving a large bottle with a narrow neck filled with cone-shaped objects to which strings have been attached.[2] Two people holding the strings need to pull the objects out of the bottle, but their likelihood of succeeding depends on the rules of the game. If they are competing to see which one can yank an object out first, their cones will create a traffic jam in the neck of the bottle, and there will be a stalemate. If, on the other hand, their task is to make sure that the objects get out of the bottle within a certain amount of time, they have an incentive to coordinate their efforts by pulling the cones out one by one. Ullmann-Margalit pointed out that something quite similar occurs when a fire alarm goes off in a crowded theater. If the fire is spreading so rapidly that only a few people will get out, there will be a scramble for the exits, which may reduce the number of people who actually do make it out alive. If the fire is spreading slowly enough that everyone can get out as long as they do so in a calm and orderly fashion, then the situation is a coordination problem. But solving this coordination problem requires shared information regarding the nature of the problem (a slow-moving fire) and its solution (calmly exiting the theater).

Examples of coordination games are easy to find in everyday life. Consumers and manufacturers of electronics have to worry about the compatibility of different systems—PC versus Mac, Betamax versus VHS, HD DVD versus Blu-ray, and so on. No consumer wants to invest in a system that will be obsolete tomorrow because most other consumers have opted for the other one. No manufacturer wants to make products that no one will buy because a consensus has formed around a competing system. During a recession, companies may hesitate to expand production and employ more workers unless they have reason to believe that others are doing the same.[3] People considering different political candidates face a similar dilemma: Is there a enough of a consensus forming around my favorite candidate that I am safe throwing my support behind her, or should I set my favorite candidate aside and give my money and endorsement to someone I like less but who has a better chance of beating the third candidate, whom I really can't stand? Knowing what to do in these situations requires some knowledge of what others are likely to do, whether it's buy Blu-ray discs or support the underdog in an election. Following the surprisingly close 2009 mayoral election in New York City in which the incumbent, billionaire Michael Bloomberg, defeated challenger Bill Thompson, backers of the losing candidate realized in retrospect that their potential supporters had faced such a dilemma. Said New York State Senator Eric T. Schneiderman, "Bill Thompson was always closer *than people thought*, and on our side, if people had been behind him more, there would have been more checks, more endorsements, more attention, and that might have made the difference."[4] Thompson's problem was not a lack of supporters, but rather a lack of common knowledge among his supporters that there were so many of them and that, as a result, their candidate actually had a chance of winning

Although coordination problems can be solved, they do not necessarily have a best or optimal solution. Any solution may be better than none, and many different solutions may be possible. Which one actually does the job is often more a matter of history and culture than rational calculation or deliberate planning. At Newark Liberty International Airport, for example, someone apparently thought that the common coordination problem of where to meet arriving passengers could be solved by putting up a sign saying "Meeting Place" in an area near the baggage carousels. But, as you might imagine, no one ever meets anyone there. Because the

would-be "Meeting Place" is located in an area that few people even notice, those few of us who know that it exists don't imagine that anyone
else is aware of it. Furthermore, the solution that works for some people
might not even occur to others. Grand Central might have seemed like an
obvious meeting place to people in New Haven, but, as Schelling himself
noted, that is certainly due to the fact that commuter trains from New
Haven arrive at Grand Central. Had Schelling instead asked people in
New Jersey or on Long Island, the most common response would likely
not have been Grand Central, but rather Penn Station, because that is
where their commuter trains arrive in New York. As New Jersey residents,
we have been to Grand Central only a few times, but we pass through
Penn Station almost every time we visit the city. Assuming that we were
trying to meet someone else from New Jersey, Penn Station would be a
much more obvious choice to us than Grand Central. And that illustrates
yet another important aspect of solutions to coordination problems: it is
not enough to have shared information; the parties involved also must
know that they all share the same information. To return to the theater
with the fire alarm, in addition to knowledge regarding the nature of the
problem and its solution, the theatergoers also need to know that everyone else in the theater understands the situation, too. If even a small number of people mistake a slow-moving fire for a fast-moving one and then
rush to the exits in a panic, the coordination problem will not be solved.
However, the same will happen if people who *do* correctly understand the
situation do not trust others in the theater to understand it, too. Common
knowledge—including common knowledge that there *is* common knowledge—is the key to solving coordination problems.[5]

The whites of their eyes: From joint attention to "Theory of Mind"

Legend has it that eighteenth-century military officers sometimes admon-
ished their troops to hold their fire until they could see the whites of their
enemies' eyes. This is most famously part of the legend of the Battle of
Bunker Hill in the American Revolution, but the same instruction appears to have been given at other battles, as well. That rule of thumb
helped the troops coordinate fire from their mostly inaccurate firearms.
But the role of the whites of our eyes—or sclera, to use the technical
term—in coordinating behavior may go back far beyond the invention of
gunpowder. Indeed, our white sclera may be just one of many adaptations to the problem of coordinating our behavior with others'.

Humans are the only primate species with white sclera. Most other
primates have sclera that are colored to match the skin around them. Due

to the way that our eyes are elongated horizontally, our sclera are also much more exposed than those of other primates, framing the colored iris in a pool of whiteness. When scanning a visual field, humans are also more likely than nonhuman primates to move just their eyes, rather than their whole heads.[6] The morphology of the human eye makes it easy to tell where someone is looking, and that may be the whole reason why our unique eye morphology evolved. Michael Tomasello and his colleagues have shown in numerous studies that humans are much better than nonhuman primates at following each other's gaze. Furthermore, while even adult chimpanzees are not particularly good at gaze following, the ability appears in humans while they are still very young.[7] Tomasello has argued that our ability to understand what someone else is looking at leads to an important first step toward the coordination of social behavior: joint attention. Joint attention simply means that two individuals are looking at the same thing. One step beyond joint attention is shared attention, in which two individuals are paying attention to the same thing while also being aware that they are doing so. From here we can move to shared intentionality, in which two individuals come to share a common intention, or goal. Because it allows people to develop different, complementary roles as they work toward their shared goal, shared intentionality is a crucial step toward the role differentiation that typifies social institutions and, thus, a crucial step toward the high degree of coordinated and cooperative social behaviors we see in our species. The attention humans pay to gaze direction and the ease with which we both follow others' gaze and use it to infer intentionality reflects the power of gaze direction to create, in a very simple and straightforward way, the common knowledge necessary for social coordination.[8]

An additional step in the evolution of human cognition that greatly enhanced our ability to coordinate our social behaviors is the development of "Theory of Mind" or "mentalizing." Although it sounds like part of a nightclub act, mentalizing is something with which we are all familiar and that most of us are very good at: understanding what probably is going on in someone else's head. "Intersubjectivity" is a closely related concept favored by some psychologists and psychiatrists.[9] Having a theory of what is happening in someone else's mind can be as simple as understanding that someone else may have knowledge that we do not have or that we may have knowledge that others do not have. Psychologists have used a variety of tests to explore Theory of Mind abilities and intersubjectivity. One such test is called "the Charlie task." The subject, usually a child, is shown a face, called Charlie, surrounded by four different kinds of candy. Charlie's gaze is clearly directed at only one type of candy. The subject is asked what kind of candy Charlie wants. The "Sally-Anne test" is a bit more complex than the Charlie task. The experimenter begins with two dolls, one called Sally and one called Anne. With Anne present, Sally puts a marble or other object in a basket. Sally then leaves, and Anne moves the marble to a box. The subject, again usually a child, is asked where Sally will look for her marble when she returns.

Most people over the age of about four pass these tests with flying colors, easily identifying the candy Charlie wants and realizing that Anne will look in the wrong place for her marble. But some individuals over the age of four with otherwise normal intelligence, such as those with particular kinds of brain damage and autism-spectrum disorders, often fail these kinds of tests. They can see what Charlie is looking at, but they cannot use that information to figure out what he wants. Unable to create mental models of others' mental states, they cannot imagine that Anne does not know as much as they do about the location of the marble. This suggests that our Theory of Mind abilities are due to selection in favor of sophisticated social cognition specifically rather than selection in favor of high intelligence in general. This may reflect how important it was to our ancestors to be able to coordinate their behavior with that of others.[10] Primatologist Sarah Blaffer Hrdy recently made the intriguing suggestion that the evolutionary roots of human intersubjectivity may lie in one of our ancestors' reproductive innovations: cooperative breeding. To a much greater extent than is found among most other primates, human infants are cared for not only by their own mothers but also by others—fathers, grandparents, aunts, uncles, and so on. Hrdy argues that such alloparenting would have favored infants with the ability to develop a sense of what is going on not only in their mothers' minds but also in those of other caregivers.[11] This selection pressure in favor of mind-reading abilities during infancy could have set the stage for the coordination of other social behaviors later in life. The result is our ability not only to know that others may know other things than we do, but also to know that they may have a theory of what is going on in our mind, too, and that that may include a theory of what's going on in their mind: "I know that you know that I know that you know. . . ." This provides the cognitive foundation not only for common knowledge but also for common metaknowledge, that is, common knowledge that there is common knowledge.

The idea that selection favored those of our ancestors who were best able to coordinate their behavior with others may explain why we do so many things simply for the pleasure of coordinating our actions with those of other people. We dance together, sing and play music together, participate on sports teams together, march together, and so on. Sometimes we combine several of these activities into spectacular displays, such as the elaborate demonstrations of coordinated musicianship and group gymnastics put together for halftime shows at American football games, which are themselves sandwiched between displays of coordinated athleticism and group-group competition. Of course, marching bands were first associated not with football but with the military, and marching remains an important part of military training even though it is no longer routine on the battlefield. Such displays may help groups monitor their own ability to coordinate their behaviors while also displaying that ability to others.[12] Recent research has shown that people who engage in simple coordinated behaviors, such as walking in step and singing together, subsequently behave more cooperatively in experimental games.[13] Our internal reward system may even be configured in such a way that coordinating one's behavior with that of others becomes its own reward: a recent study showed that rowers who work out together in synchrony experience greater releases of endorphins, those natural opioids responsible for "runner's high," than rowers who work out alone.[14]

From sculling to rowing: How collective action dilemmas can become coordination problems

The kind of rowing studied in that experiment is known as sculling. Sculls are short-handled oars that are usually used in pairs. Evolutionary biologists John Maynard Smith and Eörs Szathmáry pointed out that sculling provides an interesting way to recast the Prisoner's Dilemma game.[15] The Sculling Game involves two people in boat, one seated in front of the other, each equipped with a pair of sculls. If they both pull on their oars, they get to their destination quickly but they also both pay half of the trip's energetic costs. If neither pulls, they go nowhere but they also expend no energy. If one pulls and the other does not, the lazy individual gets the benefit of movement but does not pay the cost of rowing, while the one that pulls gets the same benefit but also pays the entire energetic cost. Although the descriptive scenario is different, the payoff structure is the same as the Prisoner's Dilemma (see Box 4.2). But what if sculls are not the only kind of oar available? In a row boat, for example, each rower has only one oar. With two people in such a boat, seated side-by-side, we have a coordination problem that Maynard Smith and Szathmáry called the Rowing Game. David Hume long ago anticipated the Rowing Game: "Two men, who pull the oars of a boat, do it by an agreement or convention, tho' they have never given promises to each other."[16] If neither rows, they go nowhere, but neither pays any energetic cost. If one rows they also go nowhere (except, perhaps, in circles), but one of them incurs an energetic cost. It is only if they both row that they get the benefit of movement toward their destination.

Reconfiguring the boat is one way to turn a Prisoner's Dilemma into a coordination problem, but the same thing can be accomplished on dry land. As we saw in chapter 4, Robert Axelrod turned the Prisoner's Dilemma into a coordination problem—and thus fostered cooperation—by making it an iterated game. When players can expect to encounter one another in the future, they have an incentive to coordinate their choices so that they remain in the "cooperate/cooperate" box as long as possible. Another way to transform the game is with shared norms that transform the payoffs. Imagine that, in addition to their agreement not to talk to the police, the two criminals had friends on the outside who could be relied upon to enforce their code of silence. Talking to the police would be met not with a light prison term or a get-out-of-jail-free card, but with death. Common knowledge regarding the code of silence turns the dilemma into a coordination game because now both players have matching rather than conflicting payoffs for talking and not talking. Shared norms, such as a code of silence, thus have the ability to transform Prisoner's Dilemmas and other conflict-of-interest situations into coordination games.[17] This may be easiest to accomplish in small, stable groups because they are more likely to share common knowledge than large, unstable ones.[18]

In one way, coordination problems are actually more fundamental, more basic, than the Prisoner's Dilemma or other collective action dilemmas. As legal scholar Richard H. McAdams has pointed out, the standard Prisoner's Dilemma presupposes that a coordination problem has already been overcome: everyone involved agrees on what exactly constitutes "cooperation" and "defection" in a particular situation.[19] If there is no agreement on this point, one party might give or do either more or less than the other party expects for "cooperation," or "cooperation" means something qualitatively different to one party than to the other. Once again, common knowledge is the key. When people fail to agree on what constitutes cooperation, it is often the result of a lack of shared background and poor communication. We remember a minor example of this from the early days of our fieldwork in Kenya. As we prepared for a week-long trip to visit settlements that were far from any road, our guides asked us to bring "tea" (*chai*) to give as a gift to our hosts wherever we happened to stay. So we did, or at least we thought we did: we brought several bags of tea leaves. But it turns out that "chai" means more than just tea leaves (*majani*). It also means sugar, which we did not think to bring. By failing to bring both components of chai, we had accidentally failed to live up to our part of the agreement. The fact that such misunderstandings are common when people from different cultural backgrounds and limited abilities to communicate with each other interact highlights the importance of common knowledge in solving coordination problems.

In the real world, transforming collective action dilemmas into coordination problems is often a key to solving them. Political scientist Dennis

Chong demonstrated this for the U.S. civil rights movement, but his take-home lessons apply to collective action more generally. As described in earlier chapters, Chong's historical analysis of the civil rights movement documented the importance of solidary and expressive benefits to individual activists. Normally in a collective action dilemma we would expect a potential activist to prefer to free ride, but solidary, expressive, and reputational benefits might make it preferable to contribute to the cause. But there is also a sense in which the very existence of solidary and expressive benefits depends upon yet another coordination game: the existence of a group of people who both share enthusiasm for the same cause and who know that they share that enthusiasm. Without a group of like-minded people with whom one can feel solidarity and among whom one can garner a reputation, solidary and reputational benefits cannot exist. Reputational benefits are especially fragile. As Chong pointed out with reference to the civil rights movement, "It is appropriate to ask how prevalent expressive behavior would be in the absence of an audience or a public."[20]

Stag Hunts and Battles of the Sexes: Trust and conflict in coordination games

Just as Maynard Smith and Szathmáry's Sculling Game is a recasting of the Prisoner's Dilemma, their rowing game is actually a recasting of a much older game: the Stag Hunt. The name comes from the work of another figure from the Enlightenment, Jean-Jacques Rousseau: "Were it a matter of catching a deer, everyone was quite aware that he must faithfully keep to his post in order to achieve this purpose; but if a hare happened to pass within reach of one of them, no doubt he would have pursued it without giving it a second thought, and that, having obtained his prey, he cared very little about causing his companions to miss theirs."[21] Game theorists have formalized Rousseau's scenario, with the highest payoffs going to two hunters who coordinate their efforts to hunt a stag (equivalent to both rowers pulling on their oars), medium payoffs to individual hunters who pursue hares on their own (equivalent to both rowers doing nothing), and low payoffs to those who pursue a stag with no help from the other hunter (equivalent to one rower pulling on his oar while the other does nothing). Pursuing a stag is the best solution for both hunters provided that they are certain that the other hunter will also focus his efforts on stag hunting and not allow himself to be drawn away by a tempting hare. Failing such trust between the two hunters, hunting hare individually is the next-best strategy for both of them because hunting stag alone is a futile waste of effort.

The important role that trust plays in Stag Hunt–type games is why they are often called "trust dilemmas" and "assurance games." The trick to solving them is to create not only a pool of common knowledge about the best course of action for all concerned but also a sense of confidence that all parties will indeed stick to their commitment to pursue the goal that is best for all. An iterated Prisoner's Dilemma creates one example of an assurance game: can both players trust each other to keep choosing "cooperate" rather than "defect"? Dennis Chong's study of the U.S. civil rights movement once again provides a good real-world example. No one wants to invest in a lost cause, not only because there's no ultimate payoff but also because side payoffs are reduced. If the U.S. civil rights movement had failed, activists would have failed to gain their ultimate goal of civil rights for African Americans, but the solidary, expressive, and reputational benefits they may have garnered along the way also would have been greatly diminished.

It thus became of paramount importance for the leaders of the movement to convince potential activists that the movement had a chance. This is the assurance game. If the leaders succeed in convincing potential activists that the movement is likely to succeed, then they will succeed in attracting activists to the cause. Thus begins a feedback loop that, when successful, creates the mass movements that put pressure on governments to change (and in the case of revolutions, the pressure that causes governments to fall). As Chong and others have documented, the Southern Christian Leadership Conference did exactly this—built step by step, success by success, building trust and assurance along the way, rather than trying to be everywhere at once.[22]

Coordination games need not be entirely free of conflicts of interest. Take, for example, a husband and wife who share the common goal of spending the evening together. But the man would prefer to spend the evening at the football stadium while the wife would prefer the ballet. If they both go to either event, then they both win in the sense that they both get to achieve their highest goal of spending time with the each other. But if they go to the stadium, then the husband gets an additional payoff, as does the wife if they end up at the ballet. This is the scenario behind a classic coordination game called Battle of the Sexes. Solving it, as with all coordination problems, is a matter of common knowledge. If they cannot communicate, they face a real dilemma. If they can communicate, then they have a variety of options. One rather cynical way to generate common knowledge would be for one to telephone the other, announce very quickly where they will be meeting, and hang up. Or they could agree to take turns choosing where to spend their evenings. Or they could leave it up to random chance by, say, flipping a coin. Or they might agree that some other factor not included in the game scenario changes their payoffs so much that they both rank football and the ballet equally. For example, if football is played outdoors and the weather is bad, even the husband may prefer the ballet. Any arrangement that generates common knowledge can help them achieve their shared and greatest goal of spending time together.

Among both voters and political actors in Washington, D.C., a real-life Battle of the Sexes goes on virtually every day. Consider, for example, the issue of "strategic voting" in elections. Strategic voting occurs when an individual votes for a candidate who is not his or her first choice because the less-liked candidate has a chance of winning but the best-liked candidate does not. In the United States, we are probably most familiar with this phenomenon in the case of third-party presidential candidates. People who support the third-party candidate are often exhorted to not "waste" their votes and to instead help their second-ranked candidate to win the election. Avoiding this "waste" is a form of strategic voting—the individual is voting for a candidate other than the one the individual most prefers.[23] Virtually all voting is in some sense strategic, however. Imagine what would happen if there were no preprinted ballots in elections, and instead each individual simply voted for the candidate that he or she most preferred. If most people voted "true" but just a few savvy people voted strategically, the strategic voters could get their candidate elected with very few votes. The strategic voters would cooperate on one candidate while the true voters each voted for their own best friends. What we see by taking this example to the extreme is that the issue is less about voting "true" than it is about equality of information. Ballots with candidate names listed help level the informational playing field and allow individuals to work together to aggregate their votes in a direction that they prefer. Most voters select candidates listed on the ballot—rather than some other preferred candidate—because they know that listed candidates have a much better chance of winning the election. By picking candidates listed on the ballot, these voters are voting strategically. They are also cooperating.

Agenda setting in Washington also fits the Battle of the Sexes framework. More than eight thousand bills are introduced each session, only about four hundred become law, and any given political actor can put time and effort into only a few. In this coordination game, it is in the best interest of each interest group and member of Congress to be working on the same issue that everyone else is working on even if it is not the issue that the interest group or member of Congress cares about most. The central question is which issues will attract enough actors to make it worthwhile to spend time and effort on them. The situation is complicated because not every policy maker or every interest group would rank a given issue at the same level. For some it might be the most important policy issue to be addressed; for others it is a minor issue that they don't necessarily oppose, but that is not a central concern. The few who feel most intensely may lobby even when the issue is going nowhere. But others will not join in unless convinced that the issue has a chance. When Beth and a group of colleagues interviewed Washington lobbyists about ninety-eight randomly selected policy issues, they were repeatedly told that the most difficult obstacle they faced was simply trying to get other interest groups and government officials to pay attention to their issue.[24]

Anti-coordination games

An even more conflict-filled scenario is imagined in the Game of Chicken. The scenario will seem familiar to anyone who has seen the movie *Rebel Without a Cause*, in which two teenage boys drive their cars in a "chicken race" toward a cliff, the loser being the one who jumps out first. In the Game of Chicken, the cars are speeding toward each other rather than toward a cliff, and the loser is the one who swerves. The best outcome for each driver is for the other driver to swerve. The next best outcome for each is for both of them to swerve, in which case they remain of equal status (and alive). The third best outcome is to be the one who swerves while the other driver goes straight. The worst outcome for both of them is to be involved in a head-on crash. Because one possible result of the two players doing the same thing is disastrous, this is sometimes called an "anti-coordination" game. Variants of the Game of Chicken abound. We introduced one in chapter 2: the Snowdrift Game. Two cars are stuck in a snowdrift in such a way that digging out one car will free them both. Just as the possibility of dying in a collision can make swerving worth the cost to one's reputation in the Game of Chicken, digging in the Snowdrift Game while the other driver sits in his warm car may be worth the resulting physical exhaustion and emotional aggravation if one needs badly enough to get somewhere.

Another version of Chicken is called Hawk-Dove. Like the Rowing and Sculling Games, it was first introduced by John Maynard Smith.[25] Two animals both want a resource. One strategy (Hawk) is to fight for it. But fighting is costly, so if they both fight, they both pay the cost of fighting and end up having to split the resource with the other party. Another strategy (Dove) is not to fight. If both play Dove, then they simply split the resource. If one plays Hawk while the other plays Dove, Hawk wins, and neither pays the cost of fighting. As is generally the case with coordination problems, one way to solve this dilemma is for there to be a convention that helps determine which party plays Hawk and which plays

Dove. Such a convention is enshrined in a third strategy, labeled "Bourgeois": if you already possess the resource, play Hawk; if you do not, play Dove. Assuming that individuals that follow each of the three strategies are equally likely to own a resource at the outset, Bourgeois does well compared to Hawk because Bourgeois-Hawk encounters, unlike Hawk-Hawk encounters, involve fighting only if Bourgeois is the owner. Similarly, Bourgeois does well compared to Dove because Bourgeois defends its resources while Dove does not. Bourgeois players do particularly well in a world full of individuals like themselves because they avoid wasteful fights while also not abandoning their resources whenever challenged by another.

Evidence from animal behavior studies indicates that many species do indeed follow the bourgeois convention, with "owners" of resources nearly always keeping possession of them when challenged by conspecifics. Although we normally think of butterflies as gentle creatures, it turns out that they can actually be quite scrappy. Nick Davies studied speckled wood butterflies (*Panarge aegeria*). Because females of this species are attracted to sunny spots on the ground, males compete with each other to monopolize them. Because sunny patches come and go as the day goes by, there are always new patches to own and male butterflies who find themselves without patches, searching for new ones. In every contest over ownership of a sunny patch that Davies observed, the current owner prevailed. When he tricked two males into thinking that they owned the same patch, both responded with a hawkish strategy, and the ensuing fights lasted ten times longer than they did when ownership was unambiguous.[26] The success of the bourgeois strategy shows that, even in situations steeped in conflict, common knowledge can be crucial to the avoidance of disastrous outcomes.

Solving coordination problems by creating common knowledge

The bourgeois strategy is an example of a convention, and conventions are keys to solving coordination problems. Human social life is replete with conventions that help us coordinate our behaviors with others. If we like to eat our meals with others, when should we do so? Conventional mealtimes help solve that problem. One can meet one's friends at lunch because "lunch" refers to a conventional mealtime. If we like to get together with far-flung family members, when should we do so? Holidays are conventions that help solve that problem. This helps explain why even nonbelievers celebrate some religious holidays: they are good opportunities to get together with family and friends.

Conventions like mealtimes and holidays work because they are common knowledge. If only one person knows about the convention, then it does not really qualify as a convention because it will be unable to help coordinate social interactions. So people need to know that there is a convention. But they need more than that. They also need to know that others know that there is a convention. If everyone knows the same convention but no one knows that everyone else also knows it, the result will be same as if none of them know it because none of them will have any reason to behave according to the convention.

Solving coordination problems thus requires not just common knowledge but also a sort of shared metaknowledge that there is common knowledge.[27] How such metaknowledge is generated is not always clear. Indeed, in many cases it may never be generated, and coordination problems may remain unsolved. Sometimes, solutions to coordination problems are more likely than others to become conventions simply because they come more easily to mind. For example, Schelling asked forty-two people to pick "heads" or "tails." If they and a partner picked the same one, they won. Thirty-six picked "heads," probably because we always say "heads or tails," not "tails or heads," when flipping a coin. In Schelling's terminology, "heads," like Grand Central Terminal, is a prominent or salient focal point solution to the coordination problem. But when Lee tried this with students in class, he got the opposite result—most of them chose tails. When he asked them why, the answer was simple: "Tails never fails!" Apparently this phrase has become widespread during the forty years since Schelling conducted his informal study and has created a new focal point solution to this coordination problem.

Another of Schelling's experiments involved a grid of sixteen squares. If a subject and his or her partner both picked the same square, they won. Most people picked the square in the upper-left-hand corner. But what if Schelling had asked Arabic speakers, who read their script from right to left? Solutions to coordination problems are also not always the "best" solution possible according to all criteria. This is true even when conventions are deliberately designed. The great statistician and biologist Ronald A. Fisher, for example, established the convention that a scientific finding is "statistically significant" (though not necessarily important) if there is less than a 5 percent chance that it occurred by chance. Fisher actually would have preferred a percentage value corresponding to a point on a normal curve two standard deviations from the mean, but the resulting percent would have been hard to remember. He chose instead to leave 5 percent in the tail of the curve, which corresponds not to two standard deviations above the mean but rather 1.96 standard deviations. Fisher deemed the salience of "5 percent" worth the sacrifice of some precision, and scientists have followed his lead ever since.[28]

Economist Michael Chwe has explored the ways in which people arrive at common knowledge. He argues that one of the important functions of rituals is to generate common knowledge. Chwe notes that many such ceremonies take place in the round, enabling everyone to see everyone else and thus fostering the development of both common knowledge and common metaknowledge. In the pueblo communities of the southwestern United States, for example, ceremonies are held in circular structures called *kivas*, with the structure itself helping to create a sense of community and common knowledge. Elsewhere, amphitheaters have a similar effect, giving the audience not only a view of the proceedings or performance but also of one another. It is common for such ceremonies to involve group dancing, which may further enhance the sensations of both common purpose and common knowledge. Of course, some public rituals are designed and managed by elites, and some forms of common knowledge may serve their interests more than those of the less powerful members of society. Political scientist James Scott refers to the stories told by such ceremonies as the "public transcript," which he contrasts with the "hidden transcript" that the oppressed share only amongst themselves.[29] The "public transcript" is the story about how, according to those in power, the society is supposed to work, what people are supposed to believe, and thus what common knowledge and metaknowledge they have. The Soviet Union, for example, regularly put on huge public displays of military might, designed to impress not only the rest of the world but also its own population with the Soviet state's power and permanence. While such ceremonies may indeed generate common knowledge and metaknowledge and thus help solve coordination problems, it is worth keeping in mind that some solutions to coordination problems may not serve everyone's interests equally.

Although the word "ritual" brings to mind religious and civic ceremonies, it does not refer only to such high-minded affairs. Chwe has pointed out that it can also refer to something as lowbrow as watching the Super Bowl on television. Because the Super Bowl is viewed by such a large proportion of America's television audience, the game itself, the half-time show, and the advertisements it showcases have the ability to generate common knowledge. This fact was not lost on the Apple computer corporation, which famously launched the Mac during the 1984 Super Bowl with a visually arresting sixty-second commercial that cleverly played on themes of oppression and resistance from George Orwell's *1984*. Remember that computer use is, in part, a coordination problem: using a computer that runs on the same operating system as everyone else's makes it easier to share files and software, get service, and so on. In 1984, Microsoft's MS-DOS dominated the operating system marketplace, and any computer company that used a different system faced an uphill battle. Other, competing operating systems, such as CP/M, had already bit the dust. The Super Bowl's ability to create not only widespread common knowledge that a new product existed but also widespread metaknowledge that everyone else knew about it, too, helped Apple win a place for the Mac in the personal computer market.

Hunting stags and hares in the laboratory

Experimental economists have also studied how common knowledge emerges and helps solve coordination problems. John Van Huyck and his colleagues, for example, used a tacit coordination game with a structure similar to that of the Stag Hunt Game, but with seven possible choices—represented by the integers from one to seven—rather than just two.[30] If all players choose seven, then they all receive the highest possible payoff. Choosing seven is thus the equivalent of hunting a stag. But if any one player chooses six, five, four, three, two, or one—equivalent to a series of progressively smaller hares—then he or she receives a payoff higher than that received by those who choose seven. When players cannot communicate with each other, many start off by choosing seven. However, after each round players were told what the minimum chosen value was in the previous round. Subsequently, more and more chose one until, after ten rounds of the game, almost everyone chose the safest, smallest hare: one. Van Huyck et al. then had their subjects play a different game in which every player's payoff was reduced equally: if they all chose seven, then they all received the highest possible payoff. If they all chose six, then they all received the next highest payoff, and so on. In that situation, nearly all players chose seven. But when those same players were then asked to replay the original game, choosing one once again became the most popular strategy. Van Huyck and his colleagues got the same result—rapid descent toward the smallest "hare"—when they not only told everyone the minimum value chosen during the previous round but also wrote all values chosen on the board after each round. The take-home lesson is clear: coordination can be more difficult than it looks.

But other experimental work shows that coordination can indeed occur, if conditions are right. Ananish Chaudhuri, Andrew Schotter, and Barry Sopher used the same game as Van Huyck et al. to explore how this might happen.[31] Chaudhuri et al. built some important new additions onto Van Huyck et al.'s foundation. After a group of players finished their rounds, they were asked to write down some advice for the next group of players. Because players earned more if their "descendants" earned more, they had an incentive to give good advice. Here's an example of the kind of advice that was given: "Pick 7 for crying out loud! But if there is a weirdo who picks lower, pick that number too. Pick 7!!! Trust each other it will help you too!" The researchers were particularly interested in the contrast between an experimental condition they labeled "almost common knowledge," in which the new players merely read the advice silently, and one they labeled "common knowledge," in which the advice was read aloud by the experimenter. Although players in the "almost common knowledge" condition knew that everyone else had been given a copy of the same sheet of paper containing the same advice that they had been given, they did not know how completely or carefully any of the other players had read it. In the "common knowledge" condition, in contrast, players not only heard the advice but also knew that everyone else had also heard it. The distinction between "almost common knowledge" and "common knowledge" turns out to be crucial. When subjects read their predecessors' advice to themselves, Chaudhuri et al.'s results looked a lot like Van Huyck et al.'s: Rapid descent to the choice with the lowest value. But when the experimenters read the advice aloud, the opposite occurred: most groups converged on the highest possible value. But even that result was surprisingly fragile. If even a single predecessor gave advice other than "choose seven," coordination fell apart. The lesson again is clear: to solve coordination problems, you need not only common knowledge, but also common knowledge that there is common knowledge.

A game similar to the Stag Hunt has also been used to shed light on the formation of ethnic groups. Why do people take such care to display markers of their ethnicity—hair styles, clothing styles, accents, and so on—and pay so much attention to the ethnic displays of others? Building upon classic anthropological work on ethnicity by Fredrik Barth and others,[32] Richard McElreath, Robert Boyd, and Peter J. Richerson have argued that ethnic markers serve to let people know who is playing the same game that they are playing.[33] Sending and paying attention to ethnic markers can thus make it easier to coordinate one's behaviors with those of others, even if the markers themselves (e.g., clothing styles) bear no direct relationship to the behaviors being coordinated (e.g., business deals). To explore this, McElreath et al. set up a computer simulation centered on a game that is similar to the Stag Hunt except that the payoffs to hunting "stag" or hunting "hare" are equal. All that matters is that you and the other player both agree to hunt either stag or hare. If your choices agree, you both get the highest payoff possible. If they don't agree, you both get a lower payoff. The problem for players is that they can't tell who is likely to make the same choices they make and who is not. To help solve this, McElreath et al. gave each of their virtual players one of two "marker traits," labeled 1 and 0, and a propensity to prefer interactions with other players with the same marker trait. In a simulation with two interacting populations, over time one population tended to become dominated by virtual players who paired a "1" marker trait with one of the two choices in the game, while the other population became dominated by virtual players who paired a "0" marker trait with the other choice in the game. The marker traits thus became reliable indicators of who is playing which game, just as real ethnic markers help people recognize others who share their assumptions about life, and in particular about social interactions.

If ethnic markers and other indications of group membership reveal what kind of game one is playing, then we may be in a better position to understand why people sometimes shift their identities or add new identities on to their existing ones. Anthropologist Jean Ensminger has argued that conversion to Islam in Africa was economically advantageous for many people because of the way in which it connected them with institutions involved in long-distance trade.[34] She notes that it was often Islamic traders themselves who convinced people to convert, and the expansion of Islam in many areas followed trade routes. In addition to links to existing trade routes, Islam provided a common language for commerce, systems of money and accounting, and a legal code. Ensminger notes that "[i]n social and legal terms, Islam provided a way of making outsiders, insiders." This is similar to economist Paul Seabright's observation that humans are remarkably good at finding ways—including but not limited to religion—of turning strangers into "honorary friends."[35]

In another study using a Stag Hunt–type game, Charles Efferson, Rafael Lalive, and Ernst Fehr put some meat on the bones of McElreath et al.'s simulation by having actual people play a coordination game with arbitrary markers.[36] Each player had to choose one of two behaviors, labeled A and B. If two players met who chose the same behavior, they both received a large payoff. If they chose different behaviors, they both received a small payoff. The experimenters created two subpopulations consisting of five players each. In one group, A was a much better strategy than B, though two players choosing B still did better than if one chose A and the other B. In the other group, the opposite was true: B was a better strategy than A, but A was still better than an A/B combination. To make things interesting—and to make it more difficult for an equilibrium to develop—the experimenters then randomly selected one player from each subpopulation to move from population to population during each round of the experiment. Players could also choose a shape, either a triangle or circle, to associate with their choice of A or B. During any given round of the eighty-round game, players were free to choose any combination of A or B and a triangle or a circle. Players were then given the option of playing with any random player from their group or one chosen randomly from among those who had also chosen the same shape. The trick to getting high payoffs is somehow identifying others with whom you share a preference for A or B. If players tend to link the behaviors they pick with the shapes they chose and then seek interactions with others who have chosen the same shape, then even a small association between the arbitrary marker and the behavior can create a feedback loop that strengthens the association until the marker becomes a strong indicator of a preference for A or B.

Like McElreath et al.'s simulation, Efferson et al.'s experiment helps explain such common phenomena as ethnicity and in-group favoritism. Those who share even arbitrary characteristics with you may be more likely to share nonarbitrary ones with you, as well, making it easier for you to coordinate your efforts with theirs. This relates to the idea mentioned in chapter 4 that human coalitional psychology may be a product of a coevolutionary process involving culturally defined groups and their social norms, on the one hand, and genes that influence human cognition, on the other. Because groups and their norms are so important to the success of our species, we should have evolved to be able to identify strongly with the groups to which we belong. But, because culturally defined groups are rather ephemeral, we should also have the ability to shift from group to group and pick up new social norms as we do so. The people who played Efferson et al.'s coordination game clearly displayed this ability, as did the people in Robert Kurzban's experiment who identified coalitions from cues provided by clothing and statements (see chapter 4).

From stags and hares to whales and fish: Coordination in the real world

One characteristic that coordination problems share is that they have no single solution. Which solution—if any—that people work out to a particular coordination problem depends not only on their rationality but also on their history and cultural traditions, which can be quite arbitrary. Some game theorists find coordination problems uninteresting because they cannot be solved simply through the use of logic and mathematics. To ethnographers and historians, on the other hand, such problems represent a wonderful opportunity to put their expertise to work in the study of cooperation.

The people of Lamalera, a community on the Indonesian island of Lembata, are subsistence whale hunters. These are not the commercial whalers who upset conservationists. Lamaleran whalers go to sea in small boats with paddles and woven palm sails. Lamaleran whaling boats focus on sperm whales, but they will also hunt other large marine animals, including other kinds of toothed whales, porpoises, whale sharks, and rays. A whaling boat is crewed by anywhere from eight to fourteen men. Most of them focus on rowing, but some bail the boat out, one serves as helmsman, one as the harpooner, and one as the harpooner's assistant.[37] When a boat gets near a whale, the harpooner leaps onto its back, plunges the harpoon home, and is left behind on the surface as the whale dives. The whale then pulls the boat until it gets exhausted or, on rare occasions, the boat is capsized. Sometimes, more than one boat will harpoon the same whale, but only if asked to do so by the first crew to harpoon the whale. Obviously, Lamaleran whaling is not for the faint of heart, and organizing whale hunts presents Lamalerans with something of a coordination problem.

Anthropologists Michael Alvard and David Nolin have argued that, for Lamalerans, whales are very much like the stag in the Stag Hunt Game. What's the hare? Fish. In addition to whales and other large marine animals, Lamalerans eat a great deal of fish. Like hunting hare in Rousseau's scenario, fishing requires much less coordination with others than whaling but does not produce as much food on a per capita basis. Lamaleran men therefore face a choice between fishing, which they can do alone or in small groups and which gives them control over their entire catch, and whaling. Whaling is a viable option thanks to the fact that Lamalerans have strict rules regarding the distribution of whale meat. All of those getting in the boat know what they will receive, should they be lucky enough to catch a whale. Though the rules governing whale meat distribution may be somewhat arbitrary, the fact that they exist and the fact that they are common knowledge among Lamaleran whalers reduces everyone's uncertainty about how their efforts will be rewarded. This demonstrates the crucial role that culture plays in creating the common knowledge that is needed to solve coordination problems.

In addition to the question of whether to fish individually or join the crew of a whaling boat, Lamaleran males who wish to hunt whales must also make another decision: which boat crew to join? In the absence of any sort of custom or convention, decisions about boat crews could become a bone of contention among Lamaleran men as boat owners compete for the best hunters and hunters compete with each other for slots in the best boats. This problem is solved by another convention regarding boat crew memberships: the bulk of the men in a boat belong to the same patrilineal descent group. Descent groups are common around the world. They organize people in a community into discrete categories based on descent from a common ancestor. One nice feature of descent groups for solving coordination and collective action problems is that their membership is unambiguous: you either are a member of a particular descent group, or you are not. This makes them better for solving and preventing conflicts about who has the right or obligation to contribute to a collective good than other, less well-defined groups. For example, a person's kindred is simply his or her circle of relatives. Although some societies do use kindreds to organize collective action (e.g., subsistence whalers in Alaska),[38] they are less useful than descent groups for this purpose because their boundaries are not well defined.[39] The Lamaleran convention that a boat's crew will consist primarily of members of a single lineage prevents conflict and makes cooperation more likely by creating common knowledge and shared expectations. The specificity of descent groups as a solution to a coordination problem is highlighted by the fact that descent groups do not structure all interactions among relatives. For example, when Lamaleran households share food, they do so not with regard to descent group membership, but, in keeping with kin selection theory (see chapter 2), with regard to biological relatedness.[40] Because food sharing is not a coordination problem, Lamalerans have no reason to turn to their descent group structure to find a solution to it.

Controlling pests on water mountains: A Balinese coordination problem

About a thousand kilometers west of Lamalera, on the Indonesian island of Bali, the problem is not how to hunt whales but how to distribute water. Bali is a mountainous island, and Balinese farmers have long made a living by terracing hillsides and flooding rice paddies with water controlled by a complex system of dams, weirs, and canals. The flow of water is managed in a decentralized fashion by local organizations of farmers called *subaks* and water temples that coordinate the actions of multiple subaks. Anthropologist Steve Lansing has conducted a series of studies to determine how Balinese farmers cooperate without any central authority to supervise the irrigation system.[41] In the process, Lansing has provided a wonderful example of how good ethnography can contribute to our understanding of real-world cooperation.

Although it is now clear that Bali's traditional decentralized system for distributing water works quite well, its effectiveness was not always understood. During the Green Revolution, high-yield rice varieties were substituted for traditional ones and new irrigation equipment was installed. As a result, the old system for scheduling irrigation fell apart, pest populations exploded, and harvests fell rather than increased. The genius of the traditional Balinese system is in balancing the needs of upstream and downstream farmers. Because water flows downhill, upstream farmers have a natural advantage over downstream farmers: they can capture and control the water as they wish. So why don't they simply keep the water for themselves? It turns out that downstream farmers have a different sort of leverage over the system: pests. Unlike water, pests can and do move uphill. Flooding paddies controls pests by depriving them of a place to live. Upstream farmers have an incentive to release water to downstream farmers, even though they could benefit by keeping some of that water for themselves, because the water allows the downstream farmers to flood their fields and thus control pests that might otherwise climb up the mountainsides and devastate the crops of the upstream farmers.

Lansing, together with John H. Miller of the Santa Fe Institute, has modeled this as a simple coordination game between an upstream farmer and a downstream one that includes the negative effects of pests on both farmers' crop yields if they fail to coordinate their behavior by planting at the same time and the negative effect on the yield of the downstream farmer if he plants at the same time as the upstream farmer and so must use less water than if he were to plant at a different time. It turns out that both farmers will prefer to plant and flood their fields simultaneously as long as the losses they would otherwise suffer from pests are greater than the losses the downstream farmer would suffer from having to share water.[42]

To see whether this simple game theoretical model had any validity in the real world of Balinese farmers, Lansing conducted a survey to find out what farmers at different elevations were most concerned about. Not surprisingly, water was a concern for everyone, but much more so for downstream than for upstream farmers. On the other hand, downstream farmers are much less concerned about pests than are upstream farmers. These concerns play themselves out through discussions at the subaks and water temples, leading to coordinated farming schedules up and down Bali's mountainsides and crop yields that cannot be beat by hybrid varieties of rice and advice from Western agricultural experts. The involvement of the temples may also be a key to this system's success. As we saw in chapter 4, religious signals of commitment appear to be more persuasive than nonreligious ones, and religious belief systems lead to stronger and longer-lasting cooperation and communities than secular belief systems. Like the system Lamalerans use for distributing whale meat, the Balinese system of water management has developed over a long period of time and has been passed down over many generations. Cultural traditions like these are often crucial to the development of the common knowledge that solves coordination problems.

Frames, scripts, schemata, and the coordination of everyday life

Common knowledge is crucial to even the most mundane and everyday social interactions. When you eat in a restaurant, you know what to do. You have a sort of script you can follow, one that includes things like "menu," "order," "bill," and "tip." Your server has a complementary script that includes many of the same elements. If you are experienced at living in any particular society, then you know many such scripts, and they help you maneuver successfully through your day, coordinating your behaviors with others without having to invent a bunch of new, one-time-use scripts as the day goes on. If you move to a new society, you will have to learn many new scripts. Many aspects of any society's scripts may seem arbitrary (e.g., knowing that when English speakers greet each other by saying "How do you do?" they are not really asking a question), but they help get people on the same page regarding what kind of social interaction they are engaged in and thus what is expected of them. Scripts, also known as schemata, are closely related to the idea of frames, and there is a large experimental literature on the impact that frames can have on how people think and behave. Frames, scripts, and schemata create common knowledge, making social life easier and cooperation more likely.

Lee has studied a frame that is important for the coordination of social behavior among Maasai pastoralists in East Africa. Maasai depend upon their livestock, and they have a variety of ways of lending, sharing, and giving animals to each other. Perhaps the most important is called *osotua*. Osotua's literal meaning is "umbilical cord," a metaphor that is both evocative and memorable. Robert Sugden has suggested that metaphors often play a role in the solutions to coordination problems,[43] and that is certainly the case for osotua. Osotua also supports a suggestion made by economic anthropologist Steve Gudeman that people around the world understand their economic systems through certain key metaphors.[44] Agent-based models indicate that, by pooling risk, osotua exchanges help Maasai herds survive longer.[45] As we suggested in chapter 4, the osotua norm may thus be one of the characteristics that have contributed to cultural group selection by making Maasai ethnic identity attractive to non-Maasai.

Osotua's centrality to Maasai life was noticed by Christian missionaries, who evoked the idea of the Bible as a bond between God and people by translating "testament" as "osotua." In addition to livestock, osotua partners, called *isotuatin*, may give each other gifts and services of many kinds. For example, young Maasai men are initiated into adulthood in a process that includes circumcision without anesthesia, and a common way for an osotua relationship to begin is for an initiate to ask an older man to hold his back during the ceremony. Similarly, being someone's best man at his wedding is a favor that is the starting point for many osotua relationships. However, livestock are so important to both the Maasai economy and to osotua relationships that ethnographers have often translated isotuatin as "stock-sharing partners," and "stock friends."[46] Osotua has also been translated as "peace" but in the sense of a peace treaty or peaceful bond between former enemies rather than a general state of peacefulness, for which the Maasai have another term (*eseriani*).[47] Ethnographer Alan Jacobs made the important observation that because osotua in its literal sense refers only to a *human* umbilical cord, its metaphorical use emphasizes the humanness of such relationships (according to Jacobs, a nonhuman umbilical cord is called *osarikoma*).[48]

To learn more about osotua, Lee conducted a series of studies both at his field site in Kenya and at his lab in the United States.[49] He began with interviews with Maasai men about how osotua relationships work. His interviewees, ten Maasai men ranging in age from twenty-five to seventy-three, agreed on the major features of osotua relationships. Osotua relationships are started in many ways, but they usually begin with a request for a gift or a favor. Such requests arise from genuine need and are limited to the amount actually needed. Gifts given in response to such requests are given freely, but, like the requests, are limited to what is actually needed. Once osotua is established, it is pervasive in the sense that one cannot get away from it. Osotua is also eternal. Once established, it cannot be destroyed, even if the individuals who established the relationship die. In that case, it is passed on to their children.[50] Osotua does not follow a schedule. It will not go away even if much time passes between gifts. Although osotua involves a reciprocal obligation to help if asked to do so, actual osotua gifts are not necessarily reciprocal or even roughly equal over long periods of time. The flow of goods and services in a particular relationship might be mostly or entirely one-way, if that is where the need is greatest. Not all gift giving involves or results in osotua. For example, some gift giving results instead in debt. Osotua and debt are not at all the same. While osotua partners have an obligation to help each other in time of need, this is not at all the same as the debt one has when one has been lent something and must pay it back.[51] Going along with the idea that osotua gifts do not repay debt, osotua gifts are not payments at all, and it is inappropriate to use the verb "to pay" (*alak*) when referring to them. Osotua partnerships are imbued with respect, restraint, and a sense of responsibility in a way that non-osotua economic relationships are not. In the words of one interviewee, *keiroshi*: they are heavy.

Interviewees disagreed on only one point: Whether anything could end or "cut" (*adung'*) osotua. Eight said that nothing could end an osotua relationship. One said that a war could end an osotua relationship. Another said that a lie, whether told to elicit a gift (or a larger gift than actually needed) or in response to a request from an osotua partner, would end the relationship. However, he also made clear that such behavior was unthinkable. Osotua partners are expected to request only what they need and to give what is needed (though no more than that) if they are able to do so.

One of Lee's interviewees illustrated many of osotua's main features with a story about his own family. Some decades ago, his ancestor Kimbai was killed by two men from an enemy group. One of Kimbai's killers then removed his warrior's belt (*ntore*) and wore it as a trophy. After the fight, the killers visited a man from another local group and asked him for food, lodging, and medicine to treat their wounds. Unbeknownst to the visitors, their host and Kimbai were isotuatin. That man's wife recognized Kimbai's belt and deduced that the visitors had killed him. She and her husband slaughtered a sheep for fat to feed the visitors, poisoned the fat, killed the two visitors, and thus avenged Kimbai's death. This revenge killing was a form of osotua gift back to the dead Kimbai and, by extension, to his survivors. The belt was then returned to Kimbai's grandfather, and the bond of osotua has continued between the two families ever since.

Readers unfamiliar with ethnographic methods might be surprised by the relatively small number of people Lee interviewed. The high degree of consensus among his interviewees illustrates why this is a safe strategy for some aspects of culture while also revealing something important about social coordination norms. The number of people an ethnographer needs to talk to in order to document a particular culture trait depends upon how consistent members of a society are in terms of how they understand that trait. Language is the limiting case. To document a language, a linguist really needs only a single fluent native speaker. Additional speakers would certainly help, but just one is sufficient because a language's lexicon and grammatical rules need to be widely shared in a community of speakers in order for language to do its communicative job. Because language is primarily a tool for social coordination, speakers of a language also have no incentive to deceive each other about its rules and learners of a language have no reason to be skeptical about them. If a native Spanish speaker teaches you that the word for "head" is *cabeza*, it makes no sense to argue otherwise. Social coordination norms, such as Maasai rules regarding osotua relationships, are similar: they work to coordinate social behavior only if everyone understands them in more or less the same way. This illustrates yet again the role of common knowledge in coordinating social behavior.

To explore the osotua norm's impact on behavior, Lee had one hundred Maasai men play the Trust Game (see Box 1.1), with half the games framed by the word "osotua." In the version of the Trust Game that Lee chose for this study, the two players remain anonymous to each other throughout the game, and each is given equal amounts of cash. The first player can give none, some, or all of his endowment to the second player. The experimenter triples that amount and then passes it on to the second player. The second player can then give some, none, or all of the funds in his control to the first player. Lee and his Maasai assistant gave all players standard instructions in Maa, the language of the Maasai, about how to play the game. Half of the fifty games were played with no framing beyond the instructions themselves. The other half were played with a single additional framing sentence: "This is an osotua game." That minimal framing resulted in several contrasts between osotua-framed games and unframed games. In keeping with the emphasis in osotua relationships on restraint, respect, and responsibility, amounts given by both players as well as the amounts that first players expected to receive in return were all lower in the framed than in the unframed games. In games played without osotua framing, a positive correlation was found between amounts given and amounts expected in return, suggesting that players were invoking such common principles of exchange as trust, investment, and tit-for-tat reciprocity. In the osotua-framed games, in contrast, no relationship was found between amounts given and amounts expected in return. In osotua-framed games but not in unframed games, amounts given by the first player and proportional amounts returned by the second player were *negatively* correlated, suggesting that the osotua framing shifts game play away from the logic of investment and toward the mutual obligation of osotua partners to respond to one another's genuine needs, but only with what is genuinely needed.

The ability of such minimal framing to shape behavior was striking. This may reflect the fact that social coordination norms, including the rules that govern osotua relationships, can be effective in coordinating social behavior only if people allow their behavior to be shaped by them. This raises the possibility that human genes and culture have coevolved with regard to our susceptibility to social coordination norms. Given how important it is for members of our species to be able to coordinate their behaviors with those of others, those among our ancestors who noticed social coordination norms and allowed their behavior to be shaped by them may have survived and reproduced at higher rates than others who were not so endowed. If this is true, then it should be fairly easy to influence people's behavior as long as you try to do so with something that is easily recognized as a social coordination norm. To test this idea, Lee and his student Helen Wasielewski explored the impact of the osotua norm on the behavior of American subjects with only minimal exposure to the concept. To get a baseline regarding how Americans play the Trust Game, they had seventy players read a text about meteorology, take a short quiz about it, and then play Trust Games presented to them with no further framing. The meteorology text was simply a dummy frame that Lee and Helen did not think would trigger players' social cognition mechanisms and that they included just in case reading some text and taking a quiz about it might have an unforeseen effect on how people play the Trust Game. Seventy other players read a short description of Maasai culture and the osotua concept and then played a Trust Game that was presented to them with no further framing. A third and final group of seventy players read the same description of Maasai culture and osotua and then played a game labeled "the osotua game." Simply reading about the Maasai and osotua seems to have had a priming effect,[52] making players feel warm and fuzzy and making them both more generous and more trusting compared to players in the dummy frame. Players who both read about the Maasai and osotua and who were told that they were playing "the osotua game," in contrast, replicated in almost every way that of the Maasai, with lower amounts being given and expected in return than in the games played after reading about the Maasai but with no further framing. This suggests that even unfamiliar social norms can have rapid and strong effects on behavior. It also indicates that the description of the osotua norm provided to the American players, which was essentially the same as the description given here, was accurate and detailed enough to result in behaviors that corresponded closely with those seen in people who had learned about the osotua norm simply by growing up as Maasai.

Coordination, evolution, and emergence

Coordination is fundamental to cooperation. Before the collective action dilemma can be solved, coordination must occur: people must share common knowledge about the nature of the dilemma and their options for solving it. Although the evolutionary study of cooperation has focused on altruism, the real key to understanding the evolution of human cooperativeness may be coordination. Our species' great success may lie less in our willingness to behave altruistically toward one another than in our ability to share one another's intentions.

Humans are able to create the common knowledge with which they solve coordination problems thanks to their evolutionary heritage. Thanks to our white sclera, we can easily see what others are looking at. Joint attention then sets the stage shared intentionality, which leads, in turn, to the ability to create complex and accurate theories in one's own mind regarding what is going on in someone else's. These cognitive adaptations create room for shared understandings that solve coordination problems in everyday life by providing guidelines for social behavior.

The focal points, norms, and conventions that help coordinate social behavior sometimes have fairly obvious origins, such as the powerful umbilical cord metaphor of osotua or the focality and prominence of Grand Central Terminal. Others have more mysterious origins. Why do we all drive on one side of the road but not the other? How did money, an essential tool for coordinating behavior in a complex economy, first arise? Answering these kinds of questions requires a different tool kit and one important concept: emergence.

Box 6.1. Coordination Games

Here are some of the more important two-person coordination games. The first number in each box is the row player's payoff. The second number is the column player's payoff.

		Rowing Game version ⟶	Pull on oars	Don't pull on oars
Rowing Game version ↓	Stag Hunt version ⟶	Hunt stag	Hunt hare
Pull on oars	Hunt stag	7,7	0,5
Don't pull on oars	Hunt hare	5,0	5,5

Figure 6.1. The Rowing, Stag Hunt, or Assurance Games

		Wife	
		Football	Ballet
Husband	Football	5,3	1,1
	Ballet	1,1	3,5

Figure 6.2. The Battle of the Sexes Game

		Chicken version ⟶	Swerve	Drive straight
		Hawk-Dove version ⟶	Dove	Hawk
Chicken version ↓	Hawk-Dove version ↓	Snowdrift version ⟶	Dig	Don't dig
Swerve	Dove	Dig	3,3	1,5
Drive straight	Hawk	Don't dig	5,1	0,0

Figure 6.3. The Game of Chicken

CHAPTER 7

Cooperation Emergent

IN CHAPTER 3, we asked you to imagine yourself as a flying fish. This time we have an easier assignment: Imagine waking up in a world in which all the vehicles are motorcycles. Don't worry if you don't know how to ride one. In this scenario, you do. Imagine also that there are plenty of roads to ride on, but no rules about which side of the road to favor when you meet another rider heading the other direction. Because everyone is riding a motorcycle rather than driving a car, a solution to this problem cannot emerge simply from which side of the car the steering wheel is on. The solution, if there is one, must emerge instead from interactions among riders. Because of your helmets, you cannot talk to each other. The only way information can be passed from driver to driver is through observations of others' preferences for the left and right sides of the road.

Given that there is no rule to begin with and no possibility of communicating with other drivers, you and all the other riders must first choose randomly whether to ride on the left or right side of the road. Don't worry about collisions. Instead of colliding, riders in this experiment always successfully swerve away from each other at the last minute. But swerving takes time, and so everyone tries to avoid having to do it. As you and your fellow experimental subjects encounter other riders, you keep track of which side of the road would have led to the least amount of swerving, and as you encounter other riders you choose the side that, based on your past encounters, is the best bet. Early in the experiment, when you have encountered only a few other riders, you might do a lot of switching back and forth. Your encounters are essentially samples from the population of riders, and small samples lead to a lot of uncertainty. As you ride along, you might want to weigh your observations differently, putting more stock in recent ones than in earlier ones. This is because each rider's behavior provides information not just about a random choice but rather about the tendency of the entire population, as that other rider has experienced it.

As time goes by, the amount of information represented by each rider's behavior becomes greater and greater. Fairly quickly, you and your fellow riders will reach an unspoken consensus that either the right side or the left side is the one to stick to. Why? Certainly not because the riders have gathered together and come to an agreement: that is prohibited by the ground rules of this little thought experiment. And certainly not because every rider personally encounters every other rider and then tallies up the votes for each side. That might work, but it would take a long time, and it would not be very interesting. And not because a central authority imposes a rule: the scenario includes no such authority. A consensus emerges because each rider's behavior is informed by the behaviors of all of the other riders he or she has encountered. When you encounter another rider who has already encountered five others who have themselves already encountered five others (for simplicity's sake, we'll stop there), you are in effect encountering not one but thirty-one other riders, with the more recent encounters capturing more and better information than the older ones. In such a system, information both accumulates and improves quite rapidly, and soon everyone is riding on the right or left.

The situation is a little like what they say in advertising campaigns against sexually transmitted diseases: when you have unprotected sex with one person, it's like having unprotected sex with everyone with whom that person has also had unprotected sex. Similarly, when you encounter someone socially, you are encountering not just that person but, in effect, all of the other people he or she has encountered in the past. Thus, each encounter taps into a large network of previous encounters. This networking and pooling of information can lead to the rapid generation of consensus regarding a wide variety of social norms and conventions, not just whether to drive on the right or left side of the road. As a result, even brief, anonymous encounters among strangers can generate norms and conventions that enable humans to cooperate, often in very large numbers indeed, by providing solutions to coordination games. Such norms and conventions are, in a word, emergent.

Emergence

In the previous chapter, we saw that people can coordinate their social behaviors if they have common knowledge both about how to do so and about the fact that everyone else also knows how to do so. Such common knowledge is often enshrined in norms about social behavior, such as which side of the road to drive on. Such norms can come from many sources. One possibility is for someone in authority to simply announce them. This is a routine thing for governments to do. For example, governments around the world issue regulations regarding electrical power (voltages, plug designs, etc.). As a result, consumers and manufacturers don't have to worry about whether appliances and household electrical power supplies will be compatible. But our motorcycle example illustrates another possibility: norms may emerge spontaneously and without planning as people go about their lives, interacting with one another. This process of social emergence represents one of the most exciting frontiers in the study of cooperation.

If the very word "emergence" makes you cringe, you are not alone. The concept of emergence has a very mixed history, and it has often been abused. Some have used it to refer to anything that is surprising or unexpected, but of course surprising and unexpected nonemergent phenomena happen every day.[1] Others use it in their struggle against the bogeyman of reductionism, cordoning off emergent phenomena as not only unreducible but also undeducible, unpredictable, and unexplainable.[2] At its worst, such usage of the term has veered away from science and into mysticism. One response to these abuses is to abandon the term altogether.[3] We argue instead that emergence, if carefully and narrowly defined, is an important—even essential—scientific concept.

Many good definitions of emergence are available. The first was provided by George Henry Lewes, who coined the term more than a century ago: "Thus, although each effect is the resultant of its components, the product of its factors, we cannot always trace the steps of the process, so as to see in the product the mode of operation of each factor. In the latter case, I propose to call the effect an emergent. It arises out of the combined agencies, but in a form which does not display the agents in action."[4] More recently, Jeffrey Goldstein has applied the concept to problems in business and management, using it to refer to "the arising of novel and coherent structure, patterns, and properties during the process of self-organization of complex systems."[5] Goldstein's definition has the advantage of highlighting self-organization and complexity, two concepts closely related to emergence. R. Keith Sawyer has defined emergence as "the processes whereby the global behavior of a system results from the actions and interactions of agents."[6] Sawyer's definition is similar to one given by Joshua Epstein and Robert Axtell for the phrase "emergent structures": "stable macroscopic patterns arising from the local interaction of agents."[7] Along those same lines, John Miller and Scott Page have defined emergence as "a phenomenon whereby a well-formulated aggregate behavior arises from localized, individual behavior."[8] Nicholas Christakis and James Fowler have written that "emergent properties are new attributes of a whole that arise from the interaction and interconnection of the parts."[9] There is nothing in these definitions that precludes ordinary scientific explanation. They merely highlight an interesting category of phenomena that is otherwise easy to overlook. These definitions also encompass a wide range of phenomena, including many that have little or nothing to do with cooperation. This chapter focuses on instances in which emergent phenomena do help people cooperate.

Emergence in the social sciences: A brief history

The concept of emergence was important in the social sciences long before it had a name. Although precursors to the idea of social emergence go back to the sixteenth century, three Scottish enlightenment figures— David Hume, Adam Ferguson, and Adam Smith—are usually given credit with having established the concept.[10] Hume's argument regarding the rules "by which properties, rights, and obligations are determined" has a game theoretical feel: "'Tis self-love which is their real origin; and as the self-love of one person is naturally contrary to that of another, these several interested passions are oblig'd to adjust themselves after such a manner as to concur in some system of conduct and behaviour. This system, therefore, comprehending the interest of each individual, is of course advantageous to the public; tho' it be not intended for that purpose by the inventors."[11] Ferguson pointed out that many social phenomena are "the result of human action, but not the execution of any human design."[12] Building upon Hume and Ferguson's foundation, Smith coined the phrase "invisible hand" and used it to explain how a participant in a market "intends only to his own gain" but nevertheless promotes "an end which was no part of his intention."[13]

Though usually unnamed, emergence subsequently made appearances in the works of many nineteenth- and twentieth-century social theorists. For example, Karl Marx's theories regarding the relationship between means and mode of production and between infrastructure and superstructure have emergentist qualities. In the twentieth century, chemist-turned-philosopher Michael Polanyi coined the phrase "spontaneous order" to describe emergent social phenomena.[14] Polanyi observed that science itself is a kind of spontaneous order, with progress being made despite the absence of an overall plan as individual scientists both interact with others and pursue their own research agendas. Polanyi's phrase was adopted and promoted by economist Friedrich Hayek.[15] Hayek's own most important contribution to the study of spontaneously emerging social phenomena was the observation that they typically make use of knowledge that is distributed among the actors involved rather than consolidated by a central authority, a pattern demonstrated in the motorcycle thought experiment with which we began this chapter.[16] Building upon Hume, Hayek, and others, Robert Sugden used game theory to examine how spontaneous social orders emerge and are maintained.[17]

Ideas emerging from this tradition have recently been paying off in a surprising corner of social science: the study of collective animal behavior. Because of the relatively limited ways in which most nonhumans interact, they are well suited to the study of social emergence. Although the tight coordination of behavior shown by birds flocks and fish schools often leads people to assume that they have evolved at the group level, careful models show that they actually emerge as individuals following simple, self-interested behavioral rules while trying to avoid both predators and conflicts with conspecifics.[18] Taking a page from Friedrich Hayek, bee expert Thomas Seeley has emphasized the role that dispersed information plays in coordinating the behavior of individuals in a hive.[19] Seeley argues that self-organization in biological systems will be favored by selection when information is decentralized and when individuals are unable to do the necessary computations. Information can be shared via both signals, which have been designed to convey information, and cues, which convey information without having been designed to do so. For example, the bees' waggle dance is a signal, not a cue, because it is designed to convey information about the location of food. A cloudy sky, on the other hand, is a cue that rain is likely, but it is not a signal because it was not designed to provide information. For coordinating social behavior, cues may have great advantages over signals, including simply being cost free. Although bees are famous for the waggle dance that they use to recruit each other for foraging, cues (e.g., the odor of the pollen on another bee) outnumber signals within beehives and are crucial to bees' ability to coordinate foraging, hive building, and other activities.[20] The value of cues over signals may apply to self-organization in human systems, as well. Imagine if you were to see an oncoming car displaying a right turn signal but veering to the left: are you better off believing the signal or the cue?

In the study of human societies, recent breakthroughs in the study of social networks, complexity, chaos, self-organization, and nonlinear dynamics have led to a resurgence of interest in the unplanned emergence of social phenomena. Paul Krugman summarized many of the key ideas in his short book *The Self-Organizing Economy*, drawing analogies between hurricanes and economic slumps and between embryonic development and urban sprawl.[21] Some of the most interesting work in this young tradition comes from a specialty called "generative social science," which uses agent-based models of artificial societies to explore the process of social emergence. Joshua Epstein, a leader in this area, has argued that agent-based models are well suited to answering the generativist's main question: "How could the decentralized local interactions of heterogeneous autonomous agents generate the given regularity?"[22]

In an agent-based model, a computer is used to simulate an environment with specified parameters and individuals with behavioral options, information, goals, and so on. Epstein and his colleagues have used agent-based models to explore emergence in a wide variety of social realms, including economics (e.g., retirement age norms), demography (e.g., the crash of the Anasazi population in what is now the southwestern United

States), social structure (e.g., the generation of social classes), and epidemiology. Archaeologists, unable to study societies in the flesh, have found agent-based models particularly useful.[23] But even those of us who study living people can benefit from the agent-based method. Steve Lansing, whose work on cooperation among Balinese farmers was described in the previous chapter, created a simulation to study how cooperation could emerge among farmers who share water. The question the farmer-agents face is when to plant their crops. By coordinating their cropping patterns, they can better control pests. The agents in Lansing's model begin with randomly assigned cropping schedules, and the lack of coordination leads to high losses due to pests. As each new year begins, however, each agent checks to see how successful its neighbors have been, and it mimics the best strategy it observes. Soon, cropping patterns are distributed across the landscape in groups rather than randomly, leading to virtual pest control and higher virtual crop yields. The overall pattern generated by the model closely resembles the actual pattern observed in Bali itself.[24]

Invisible hands . . .

Agent-based models are a new version of a very old idea: the invisible hand. Though widely misunderstood, Adam Smith's metaphor of the invisible hand is a useful tool for understanding how norms and conventions can spontaneously emerge from social interactions. Our motorcycle scenario is an example of an invisible-hand account of the emergence of a social norm. A much more classic and important example is Carl Menger's explanation of the origins of money. Menger was a nineteenth-century economist who, along with William Jevons and Leon Walras, developed the idea of marginal value. Although we now come across money in the form of currency that is produced, underwritten, and highly managed by governments, it is very unlikely to have originated through anyone's conscious design. Rather, Menger argued, money most likely originated spontaneously out of buyers' and sellers' interactions in a barter economy. Barter economies are inherently clumsy: to turn what you have but don't want into what you want but don't already have, you have to find someone who both wants what you have and is offering what you would prefer. One way an individual trader can deal with this awkward situation is to accept goods that he doesn't want but that he knows from previous experience are likely to be accepted by others. He can then engage in a triangular exchange, selling what he has but doesn't want to one party and obtaining what he wants but doesn't have from another. Just as anonymous encounters among motorcycle riders lead to a consensus regarding the left/right driving rules, such triangular exchanges lead to the emergence of a single good as a medium of exchange for other transactions. Money thus becomes an important institution enabling cooperation, first in small groups, and eventually among large numbers of strangers over vast expanses of space and time.[25]

Although no records were ever made of money as it originated, the limited information we can glean from ethnographies of societies with rudimentary currencies supports Menger's scenario. The development of shell money on the island of Bougainville in Melanesia is a case in point. In about 1908–9, Richard Thurnwald, an anthropologist who had studied under Menger, found that among the Buin at the southern end of the island shell "money" (he placed the word in quotes) was used only to pay bride wealth and allies and to pay restitution for murders, but not for trade, which was limited to barter. When Thurnwald returned to Buin in 1934, he found that shells had become true money, being exchanged for a wide variety of goods.[26] Ethnographers working elsewhere on the island told similar stories. In 1930 at the north end of the island shell money and tobacco were coming to be used as currency, though barter was still common.[27] Among the Siuai, who neighbor the Buin at the southern end of the island, shell money was preferred for all transactions in 1938, though barter had dominated until recently.[28]

The motorcycle scenario involves just two options (right or left). The money scenario involves a larger but still limited number of options because only a few goods will have characteristics such as portability and divisibility that lend themselves to becoming currency. But the logic of those examples applies even if options are continuous and unbounded. This is the case with prices in a market with money. Here's another thought experiment: Imagine a market in which sellers encounter only buyers, never other sellers, and buyers encounter only sellers, never other buyers. Thus, information is conveyed to sellers about the prices other sellers are offering only through the behavior of buyers: they either buy, or they don't. Information is conveyed to buyers about the prices other buyers are accepting and rejecting only through the prices being set by the sellers. Even with information exchange as limited as that, a consensus price will still emerge as sellers adjust their prices in response to perceived demand and buyers adjust their buying habits (i.e., the threshold price above which they will not buy) in response to perceived supply.[29]

. . . or just hand waving?

Talk about markets and invisible hands raises many people's hackles. Most negative responses to invisible-hand explanations arise from how they have been embraced, at least in name, by some political conservatives. But with liberals such as Paul Krugman and Marxists such as Jon Elster offering invisible-hand explanations, it is hard to maintain the idea that the method somehow belongs to conservatives.[30] Indeed, Edna Ullmann-Margalit argued persuasively that the association between the invisible hand and political conservatism is ironic because in earlier times it was seen as providing support for liberating ideas of secularism, enlightenment, and progress.[31] Friedrich Hayek even went so far as to write an essay titled "Why I Am Not a Conservative."[32] Some have speculated that the difficulty so many people have in accepting invisible-hand explanations may arise from a natural "artificer bias"[33] that makes us unable "to conceive of an effective coordination of human activities without deliberate organization by a commanding intelligence."[34] A similar bias that we mentioned in an earlier chapter, called a "hyperactive agency detection device," has been suggested as a reason for our eagerness to believe in supernatural forces.[35] Given that the cost of failing to detect agency when it is really there (e.g., assuming that the wind rather than a predator is making the bushes rustle) is likely to have been greater than the cost of attributing agency when none is really present (e.g., assuming that a predator is making the bushes rustle when it is really just the wind), such a tendency might have had real survival value for our ancestors. An awareness of such a tendency allows us to hold it at bay when working to explain both natural and social phenomena.

It is undeniable, however, that some invocations of the invisible hand amount to little more than hand waving, with important details left unspecified and beneficial outcomes somehow guaranteed for all concerned. Given such a track record, it is understandable that some scholars dismiss invisible-hand explanations outright. For example, evolutionary biologist David Sloan Wilson has called the idea "pure fiction."[36] Ironically, natural selection is itself an invisible-hand explanation. Although invisible-hand explanations are appropriate for any unplanned phenomena, they are most interesting when they help us understand something that appears to be the product of deliberate, intelligent design but that actually arose spontaneously and without planning. When Darwin wrote *On the Origin of Species*, he was doing exactly that, that is, providing an explanation for how a phenomenon that appears to have the hallmarks of planned design is actually the unplanned product of a natural process.[37] The social and the biological sciences thus share common intellectual roots in invisible-hand thinking. Both stand to benefit from a greater appreciation of invisible-hand explanations, including an understanding of their limitations.

To be useful, the invisible hand needs to be more than just a metaphor. In the late twentieth century, three philosophers of very different political stripes—Robert Nozick, Jon Elster, and Edna Ullmann-Margalit—systematized the invisible-hand method, turning it into it a qualitative but rigorous tool for social inquiry.[38] Ullmann-Margalit argued that invisible-hand reasoning's greatest value for the study of cooperation was in its ability to explain the origins of coordination norms.[39] She also saw it as occasionally useful for explaining norms that help settle conflicts of interest (e.g., Prisoners' Dilemmas), but not at all suitable for explaining norms that enforce and maintain situations of inequality. Following her lead, economists and philosophers such as H. Peyton Young, Robert Sugden, and Christina Bicchieri have used invisible-hand reasoning to develop formal analytical models of the processes by which norms and conventions that help coordinate social behavior emerge spontaneously from social interactions.[40]

Like agent-based models, invisible-hand explanations begin with ordinary people going about their ordinary lives. All we need to know about them is something about their goals (e.g., avoiding collisions) and their options for achieving them (e.g., riding on the right or left). It is usually best to assume that the actors are concerned primarily about their own interests. Although real actors may sometimes be concerned about the welfare of others, if the explanation is intended to account for a socially beneficial pattern, then assuming such a motive amounts to stacking the deck. Assuming self-interest makes the resulting explanation stronger because it does not depend upon anyone behaving altruistically. The rest of the explanation must conform to a rule of "normalcy," meaning that it should "sound like a description of the ordinary and normal course of events" and "cannot hinge on the extraordinary and the freaky."[41] Some add a final constraint that the actors must not have the final pattern in mind.[42] Others point out that it is all right if they do have such knowledge so long as they are not motivated by a desire to see it come true. Such a motivation would amount to a kind of planning and so take the explanation out of the invisible-hand category.[43]

Invisible-hand explanations often replace alternative explanations based on deliberate design. This makes them both surprising and challenging, which in turn makes them intellectually rewarding.[44] The single most important example of this is undoubtedly Darwin's explanation of the origin of species, but Menger's explanation of the origin of money also has these qualities. Because we now encounter money only in its modern, highly planned and regulated forms, most people naturally assume that it was deliberately invented at some point in the past, presumably by someone rich, powerful, and rather brilliant. Because Menger's explanation involves only ordinary people living ordinary lives and behaving in ordinary ways, it is both more plausible and more parsimonious than an explanation based on deliberate design.

Emergent social patterns: The good, the bad, and the ugly

A common misconception regarding emergent social phenomena is that they are somehow always beneficial. While some emergent social patterns do provide many benefits, others have effects that range from mildly annoying to quite undesirable. For an example of a mild annoyance, let's return to the world of driving. We have probably all found ourselves in traffic jams, crawling down the highway, expecting eventually to come across the source of the slowdown, such as an accident or stalled car, only to come clear of the jam without ever having learned its cause. Such ghost or phantom traffic jams emerge from the dynamics of driving. Even after the reason for a slowdown has long passed, the behavior of slowing down can persist by being passed from driver to driver, each of whom is doing precisely what he or she should do: slow down when the cars ahead do so. The information that there is a need to slow down lasts much longer than the actual need to do so. If the density of cars on the highway drops enough, then the chain of information transfer can be broken.[45]

For an example of a socially undesirable emergent social pattern, let's turn to Thomas Schelling's invisible-hand explanation of residential segregation.[46] Imagine a society in which there are two kinds of people—"Black" and "White," tall and short, Flemish and Walloons, Republicans and Democrats, or whatever. They live on a grid, and one characteristic they all share is a preference for having at least a few neighbors who are like them. If they have only one neighbor, then they want one in their category. If they have two, then they want at least one of them to be in their category, and so on up to the maximum possible of eight neighbors, in which case they want at least three that are in their category. That would still leave them in the minority, but Schelling showed that even such a mild preference for similar neighbors will lead to residential segregation as people move from place to place in search of neighborhoods that make them happy. The problem is that when one person moves to a neighborhood that makes her happier, that has an effect on the mix in her old neighborhood that will lead some of her former neighbors to decide to move, as well. Eventually, most people will be living in neighborhoods in which their kind predominates. While this may make individuals happier in the short run, it is easy to imagine that it could lead to further divisions that would be detrimental for the long-term well-being of both the society as a whole and its constituent individuals.[47]

Other emergent social phenomena may be helpful, but still not optimal. Driving on the right or left side of the road is arbitrary. We have a lot of experience driving in countries with both rules, and we can report that they both work just fine so long as everyone knows the rule. But, particularly when driving a vehicle with a standard transmission, we actually prefer to drive on the left, as we have done while conducting fieldwork in Kenya. The reason is that we are both right-handed, and the position of the steering wheel on the right side of the car commits our left hands to the relatively indelicate job of shifting gears and leaves our good, right hands on the steering wheel. This relates to how side-of-the-road patterns emerged in the real world. When people riding horses passed on the road, they mostly stayed to the left, thus leaving their right hand free to greet (or to beat, as they case may be). In some countries, that pattern persisted when people rode horse-drawn carts. But in the United States and elsewhere, carts and the teams of horses that pulled them grew so big that right-handed drivers preferred to keep their right hand free to handle the whip, leading them to sit on the left and to drive to the right so that they could best avoid collisions with oncoming traffic.[48]

Even an arbitrary convention that is demonstrably worse than another may persist if it helps coordinate social behavior. For example, we are using Microsoft Word to write this book. Why? Over the past three decades, each of us has used—and often preferred—a variety of other word processors. But, because moving files from one program to another often leads to problems, it is easiest simply to use the most commonly used word processor, despite its shortcomings. Our computers also have qwerty keyboards, about which a similar story can be told. Although qwerty results in slower typing than some other arrangements of keys, any standardized keyboard layout is useful because it makes it easy for people to move from one keyboard to another. Once a coordination norm is in place, it has a certain staying power because of the difficulty of reocordinating the behavior of a large number of people around another norm.

Types of emergent social phenomena

One of the problems with the concept of emergence is that emergent phenomena are so diverse. Even within a single part of science, it sometimes seems that emergent phenomena have very little in common other than the fact that they are emergent. In the social sciences, for instance, what do the Mexican wave, traffic patterns, prices, social norms, and fads have in common? Sociologist R. Keith Sawyer has cut the problem down to size by sorting emergent phenomena into types based chiefly on how long they last. First come "ephemeral emergents." As their name suggests, ephemeral emergents last only as long as the single interaction from which they emerge. Though they are short-lived, ephemeral emergents help create understandings between people, thus helping them cooperate. Much of Sawyer's own research concerns this level, focusing on how interactional frames emerge in the context of improvisational comedy, or "improv."[49] Improv is entertaining precisely because neither the audience nor the actors themselves know what interactional frame is at work in a particular sketch until it emerges from the action on stage.

Next come "stable emergents," which last for more than just one encounter. For example, a society's shared culture, including its language, emerge through interactions over long periods of time and help shape the ways people perceive themselves and their groups and the ways in with they interact with one another. In the world of improv, Sawyer has identified several stable emergents that have grown out of improvised sketches and that are retained because they have proven their worth in generating believable and entertaining performances. For example, the "Yes, and . . ." rule states that with every turn an actor should both accept what has been offered previously by another actor and add something to it that further develops the dramatic frame. The "Yes, and . . ." rule helps solve a collective action problem among the actors by encouraging them to contribute to two public goods: the creation of a dramatic frame and the advancement of the action of the sketch.

Finally, there is "social structure," a phrase that social scientists have used in a variety of ways. Sawyer uses it to refer to emergents that have gone beyond simple stability and become fixed in some stable material form, such as a society's technology and infrastructure and its institutions and written codes of law, as we saw in chapter 5.

The value of Sawyer's simple typology becomes clear when we apply it to particular cases. Consider money. Menger's explanation of money's origins focuses on how it became a stable emergent as a result of interactions among individual traders in a barter economy. In the early phases of that process, when some individuals have figured out the value of engaging in triangular exchanges, we might refer to the go-between commodity as an ephemeral emergent. After the process is completed and money becomes established and regulated, it is an aspect of social structure. However, Sawyer's typology should not be mistaken for a developmental sequence. Some ephemeral emergents may lead to stable emergents, and some stable emergents may lead to social structure, but that is not necessarily the case. Many aspects of social structure are deliberately planned, not emergent.

From curves to conventions: How mathematics helps shape cooperation

Because the emergence of norms and conventions often involves information dispersed among a population of actors, the mathematical properties of information often play a role in how emergence occurs. For example, a fundamental principle of probability and statistics known as the central limit theorem can lead dispersed agents to converge upon a common norm, convention, or behavior. The theorem states that even if the underlying distribution of a variable in the real world (in statistics jargon, in "the population") does not fit the familiar bell-shaped "normal" curve, a distribution of averages calculated from samples of that variable will be normally distributed and will converge on the actual average of the population.[50] Among animal behaviorists, the central limit theorem is seen as "the cornerstone for understanding all collective phenomena,"[51] and evidence of its importance is easy to find. For example, as each individual in a fish school or bird flock acts upon the information it receives from its neighbors, its actions have the effect of sharing an average of that information with those same neighbors, leading to a motion of the entire school or flock that incorporates information from all its members without any of its individual members having to be aware of all of the information involved.[52]

The central limit theorem also plays a role in the emergence of norms and conventions among humans. For example, in the motorcycle scenario given at the beginning of this chapter, the underlying distribution of behaviors clearly cannot be normally distributed because there are only two options—right and left. But the central limit theorem ensures that as each rider collects a sample of the behavior of other riders and uses them to guide his or her own behavior, the result will converge upon a single side of the road. The same argument can be made for the emergence of money, where the fact that there is a finite and discrete number of goods that might be used as money (in statistical terms, the categorical nature of the data) also means that the underlying distribution will be nonnormal. In the thought experiment presented above regarding prices, a normal distribution of the underlying variable is possible, but it is not necessary for a common price to emerge.

Another common way that data may be distributed, one that is decidedly nonnormal, can also contribute to the emergence of cooperation. A "power law" curve looks a little like a ski jump: very high on one side and then dropping rapidly and smoothly down to the x-axis, where it peters out asymptotically. It's called a "power law" curve because its shape can be described by an exponent. Power law curves are extraordinarily common in nature. Take gravity, for example. As we all know, the gravitational pull between two objects gets weaker as they get farther apart and stronger as they get closer together. What Isaac Newton figured out was that gravity does this not in a linear fashion, but following a power law curve, with the gravitational pull between two objects being inversely proportional to the square of the distance between them. The frequency of earthquakes of a given size follows a similar logic, which is known by geologists as the Gutenberg-Richter relationship. If a region experiences one magnitude 4 quake per year, it will also experience ten magnitude 3 quakes, one hundred magnitude 2 quakes, and so on.[53]

When you start looking for them, power law curves have a way of showing up almost everywhere, in social as well as physical phenomena.[54] The study of power laws in human affairs got going in 1949 when a linguist named George Zipf noticed something about word frequencies. In any language, the frequency of the second most common word is about half that of the most common word, the third most common word is about a third that of the most common word, and so on down the line. Looking beyond language, Zipf noticed that something very similar was true of the sizes of cities. In any country, the population of the largest city is always about twice that of the second largest city, three times that of the third largest city, and so on.[55]

A common way for power law curves to emerge in human phenomena is simply through copying. If there are a variety of things in a category, and if the likelihood that any one of them gets copied is a function of how common it is, then the more common ones will tend to get copied more than the less common ones, making them all the more common (and less common, at the other end of the spectrum) as a result. The result is a snowball effect for the more common ones. This phenomenon is known as cultural drift. Archaeologist R. Alexander Bentley has found just this kind of curve in phenomena ranging from ancient pottery styles to academic jargon and baby names.[56] Although we all certainly feel as if we are exercising our agency when we name our children, it turns out that, on the whole, we do so in a way that can largely be explained in terms of random cultural drift. Similarly, we may have strong feelings about our favorite breed of dog, but dog breeds also follow a power law curve, with a few breeds being very, very popular and many other breeds being not so popular. Drift appears to be a powerful and ubiquitous background effect in cultural transmission, though occasionally something will shift the curve away from pure drift. For example, Dalmatians suddenly became a lot more popular when, in 1985, Disney rereleased the movie *One Hundred and One Dalmatians*.[57]

Power laws, criticality, and assurance games

Power law curves are important for the study of cooperation because they describe a way in which one crucial obstacle to cooperation is overcome: a lack of assurance. As we saw in the previous chapter, coordination games are solved through the creation of common knowledge of how they are to be solved and—most crucially—common knowledge that there is common knowledge about that solution. As Michael Chwe has argued, what is needed is not simply common knowledge but common metaknowledge. The subset of coordination games known as assurance games requires an additional sort of common knowledge. Everyone involved needs to trust not only that everyone else knows about the solution but also that they will also act accordingly. In the Stag Hunt Game, for example, everyone needs to know (a) that hunting stag as a group will provide more meat than hunting hares individually, (b) that everyone else also knows this, and (c) that everyone else will indeed hunt stag rather than hares. Processes that can be described by power law curves have often led to the widespread shared assurance that is needed to solve this sort of problem.

This idea has been around for quite a while in the social sciences. Indeed, it appears to have been invented and reinvented several times. Each time this occurs, a new metaphor is invoked. Critical mass, avalanches, cascades, punctuated equilibria, threshold effects, tipping points, bandwagons, and wildfires are some of the more common, and they all capture some aspect of what happens when people—sometimes a great many people—suddenly realize that they share not only an understanding of how a common goal can be achieved but also a confidence that everyone will indeed act upon this knowledge.[58] As the "critical mass" metaphor suggests, the result is a lot like a nuclear chain reaction: seemingly without warning, enough people share the right kind of knowledge that a sudden (and sometimes literally explosive) change is possible.

Revolutions against oppressive regimes often have this kind of surprising element. No one sees them coming, and then, almost before anyone realizes what has happened, they have come and gone, leaving a new world in their wake. The French Revolution of 1789, the Russian Revolution of 1917, the Iranian Revolution of 1979, the East European Revolution of 1989, and the Arab Spring of 2011 are all good examples. Four months after the fall of the Berlin Wall, for example, fully 76 percent of East Germans polled admitted that they did not see it coming, despite the well-documented human tendency to exaggerate their foreknowledge of events after they have occurred.[59] Even people who have worked hard and waited long for regimes to collapse are often taken by surprise when their dreams finally do come true. The Ayatollah Ruhollah Khomeini, for example, did not anticipate the rapid collapse of the Shah of Iran's regime or his own transformation from exile to head of state.[60] Similarly, in early 1917 Lenin predicted that he would not live to see a revolution against the Romanovs.[61]

Economist and political scientist Timur Kuran has suggested that the element of surprise emerges from the fact that oppressive regimes go to great lengths to discourage and prevent their citizens from being open about their true political opinions. This prevents the spread of knowledge about public discontent and creates an illusion of a citizenry that is satisfied with the status quo. Kuran refers to the discrepancy between private and public preferences as "preference falsification." Yale political scientist James C. Scott captures much the same idea with his distinction between the "public transcript" that toes the official line and the "hidden transcript" that oppressed people share only among a few confidantes or when anonymity is ensured (e.g., through graffiti).[62] In addition to creating the illusion of a satisfied citizenry, such oppression also prevents people from realizing that their own revolutionary acts will be joined by more such acts. Thus, a society may come to the verge of revolution without anyone realizing it. When the illusion is shattered, the floodgates are opened, and antipathy and anger toward the regime are free to flow out. Alexis de Tocqueville said it best: "Patiently endured so long as it seemed beyond redress, a grievance comes to appear intolerable once the possibility of removing it crosses men's minds."[63]

A very similar dynamic has shown to be at work during many strikes by laborers, particularly before the advent of strong unions and collective bargaining. Discontent with compensation and working conditions may be widespread, but until an understanding of its extent also becomes widespread, no one takes any action. When this information logjam is broken, collective action often takes place with great speed and dramatic results. Sociologist Michael Biggs has emphasized the role of positive feedback in generating the level of common knowledge necessary for strikes of this kind to occur.[64] Just as a wildfire may begin with a single spark, a strike may begin with a single worker willing to express his discontent. This emboldens others, who, in turn, embolden yet others. Biggs explores this idea with data from strike waves that occurred in Chicago in the 1880s and in Paris in the 1890s.

The Chicago strikes of 1886 are legendary among labor activists for the Haymarket affair, a demonstration marred by violence and followed by a controversial trial that inspired the establishment of May Day as an international workers' holiday. Chicago was by no means the only American city to experience massive strikes that year, but the number of workers involved there was far and away the largest. The May strikes were followed by more strikes, some lasting just for a day and affecting a single firm and others having more widespread impacts. Lucky for Biggs and other modern scholars, detailed records were kept of the strikes by the U.S. Bureau of Labor Statistics. Using those data, Biggs has shown that the pattern of strikes followed a power law curve, with many strikes affecting just one or a few firms and a few affecting hundreds of firms. Biggs found similar patterns for strikes among Parisian workers in the 1890s, suggesting that the pattern observed in Chicago is not an isolated case.

Power laws and powerful lobbyists

The relevance of power laws to the study of cooperation is by no means limited to the sometimes violent worlds of revolutions and strikes. Virtually the same patterns emerge when we examine how the political agenda is set in Washington, D.C. Any given political actor can put time and effort into only a handful of the thousands of possible issues. Lobbyists thus face a difficult choice. Each one could try to influence government policy on a different issue, but doing so would probably be ineffective. The situation is complicated by the fact that not every policy maker or interest group would rank a given issue at the same level. For some, a particular issue might be the most important one to be addressed; for others it may be a minor issue that they don't necessarily oppose but that is not a central concern. But even if an issue is not one's highest priority, it makes sense to work on an issue on which others are already working because it has some chance of actually receiving attention from legislators or regulators. As we explained in the previous chapter, this situation is essentially a Battle of the Sexes game, with some preferring one outcome and others preferring another, but everyone preferring coordination to a lack of coordination.[65]

Together with several colleagues, Beth has examined this issue by interviewing both lobbyists and congressional staffers and by analyzing records of lobbying created by the Lobbying Disclosure Act of 1995.[66] The problem of attention can seem like a catch-22. An issue cannot go forward in the political process unless it receives more attention, but it cannot get more attention unless it goes forward. An issue becomes "important" and worth spending time on because others have decided that it is important and worth spending time on, and those decisions are in turn based in large part on the decisions of others before them. Despite this problem, bandwagons do form, but only around a few of the thousands of possible issues. The resulting pattern forms (yes, you guessed it) a power law curve. Out of 137 issues that Beth and her colleague Frank Baumgartner examined, only 4 of them garnered the attention of more than five hundred organizations—more than a third of all interest group activity in their sample. Most issues attracted the attention of none, one, or a few organizations.

Revolutionaries, workers, and lobbyists are not the only ones whose success depends upon them finding a popular bandwagon to join. Scientists also face a sort of coordination problem: What is considered an important question? Which new approach will lead to the next breakthrough or paradigm shift, and which one will fizzle out? Due to their diverse backgrounds, scientists who wish to understand cooperation face a particularly difficult coordination problem. In the next chapter, we have some suggestions about how they might find a salient focal point solution.

CHAPTER 8

Meeting at Grand Central

WE STARTED THIS BOOK with four vignettes, each highlighting a contrast between a situation in which cooperation did occur and one in which it did not. Let's take another look at those vignettes in light of the theories and concepts described in the intervening chapters.

Water as a common-pool resource

Fresh water is a scarce resource around the world, but particularly in arid regions such as the American West. At one time, groundwater was sufficient for the needs of the region's small population, but rapidly growing populations in recent years have led to the depletion of aquifers and the diversion of enormous amounts of water from the Colorado and other rivers. Conservation efforts have been, for the most part, sporadic and ineffective, and, for many communities in the region, a water crisis looms on the horizon.[1] In contrast, communities of farmers around the world have been successfully sharing irrigation water for many years. In some cases, such arrangements have existed for centuries. Farmers in Valencia, Spain, for example, still use rules for water distribution that were drawn up in 1435.[2]

In much of the American West, water is a contentious issue, and water distribution is highly politicized. The situation in central Arizona, located in the Sonora Desert, is particularly dire. A history of water subsidies, mainly by the federal government in the form of dams and canals, has encouraged the overuse of water by all types of users in the Phoenix area—agricultural, residential, and commercial. Despite the natural aridity of the environment in which they live, residents of the Phoenix area are largely oblivious to the problem. Swimming pools and lawns are the norm, and golf courses are commonplace. As a result, aquifers are being depleted, and water shortages are predicted for the future. The situation is exacerbated by rapid population growth that has been enabled and

encouraged by the availability of cheap, subsidized water. Efforts to encourage conservation have met with little success and much opposition.[3]

As we saw in chapter 5, common-pool resources are those with low
excludability (i.e., it is hard to prevent people from using them) and high
subtractability or rivalry (i.e., each additional consumer helps to deplete
them). For both theoretical and practical reasons, common-pool resource
management is one of the most important areas for research on cooperation. As Garrett Hardin pointed out, such resources are vulnerable to the
Tragedy of the Commons, a situation in which a resource is depleted because everyone has an incentive to use it before it is gone.[4] At one time it
was thought that there were only two ways to deal with common-pool
resources: eliminate them by dividing them and converting them to private property or use government power to ration them and overcome
conflicts among users. In central Arizona, both have been tried, and the
situation continues to deteriorate. Only a small proportion of the water
resources in the region are in private hands. Most of it is managed by a
variety of government agencies. Far from solving the problem, government management of water has essentially created the problem by subsi-
dizing water delivery and storage. Government management has also
turned water into a political issue, making it difficult to raise the price of
water or impose any lasting or effective conservation measures in the
area.

The Arizona situation contrasts with that in Valencia, Spain, a semi-
arid region where farmers share water with the help of rules their ancestors codified in 1435. The Valencia example is a literally a textbook case
of a successful common-pool resource management scheme: it is featured
in Elinor Ostrom's landmark *Governing the Commons*. Ostrom, who
based her account of the Valencia system primarily on the work of historian Thomas Glick and political scientists Arthur Maass and Raymond
Anderson, pointed out that some of the rules that were codified in 1435
had already been in use since long before the region was recaptured from
the Muslims in 1238, adding up to about a thousand years (so far) of
successful management of a common-pool resource.[5] Such remarkable
continuity reflects the fact that the Valencian system includes all of the
elements that researchers such as Ostrom have found to be necessary for
the successful management of a common-pool resource. The resource
and those who have the right to use it are both clearly defined. The rules
that are used to allocate the water are appropriate for the local circumstances and are flexible enough to adjust as those circumstances change,
such as during a drought or a period of exceptionally abundant water.
There are well-established and effective procedures for monitoring compliance with the rules, enforcing the rules, and settling disputes, including
a Tribunal de las Aguas that meets every Thursday morning. Despite the
constant temptation to cheat by drawing off more water than the rules

allow, records going back to the fifteenth century show that cheating has been very rare.

Grassroots justice in Tanzania

....

Citizens of many countries are frustrated by ineffective and corrupt police forces. Although they may occasionally take matters into their own hands on an individual and ad hoc basis, it is rare for them to organize a viable alternative to the police force. In Tanzania in the 1980s, such a rare event did occur. Members of the Sukuma ethnic group organized Sungusungu, a system of grassroots justice and vigilantism. Sungusungu was so successful that it was deputized by the Tanzanian government and imitated by other Tanzanian ethnic groups. But the imitators did not always share the success of the Sukuma. Members of the Pimbwe ethnic group, for example, attempted to form their own Sungusungu, but they eventually abandoned the effort in frustration.

Although readers of this book probably live in places where the police do their jobs effectively and the justice system works fairly well, that is not the case for millions of people around the world. For them, the police are a corrupt and unreliable force, and justice is often a rare commodity. In Tanzania, the situation grew particularly bad following a war with neighboring Uganda in 1979. Although Tanzania prevailed in the war, sending Uganda's dictator Idi Amin into exile and destroying his government, Tanzanians living near Uganda suffered from an increase in crime, particularly cattle rustling. Members of two of Tanzania's largest ethnic groups, the Sukuma and Nyamwezi, fought back by forming an organization called Sungusungu. Sungusungu can be described as either a grassroots justice organization or a vigilante force—or perhaps some of both. Through a combination of intimidation and an ability to supply services not effectively supplied by the government, Sungusungu spread rapidly and grew into large, hierarchically organized bureaucracy with good record-keeping practices, clearly defined rules and procedures, a variety of specialized roles, and an ability to track and capture wrongdoers, even over long distances. Sungusungu's secret councils, which hear allegations of wrongdoing, pronounce judgments, and determine and dispense punishment, are its most powerful component. Punishments include ostracism, fines, public humiliation, beatings, and even death. Although Sungusungu's relationship with the Tanzanian government has often been strained, it was so successful in the 1980s that it was officially deputized by the government. Deputizing came with some restrictions, such as a ban on firearms and a requirement that Sungusungu work with the Tanzanian police. Although Sungusungu has often been compared to the mafia or a vigilante organization, there is no denying that it represents a very successful attempt to overcome the collective action dilemma and provide an important public good.

Sungusungu has been so successful among the Sukuma and Nyamwezi that some of Tanzania's other ethnic groups have tried to mimic it. For example, the Pimbwe, a much smaller ethnic group in western Tanzania, attempted to create their own Sungusungu in an effort to solve a common problem: incursions by Sukuma cattle into their fields. Unlike the Sukuma and Nyamwezi Sungusungus, the Pimbwe's failed quickly, having gained the participation of fewer than 10 percent of adult men. Why were the Sukuma and Nyamwezi so successful compared to the Pimbwe? Brian Paciotti and his colleagues see the key differences as being cultural and institutional. The Sukuma are famous in Tanzania for their strong ethic of generosity. Pimbwe are known for being rather stingy. Even before Sungusungu developed, Sukuma communities had a variety of large organizations, such as dance societies, that helped create the relationships and organizational principles that made Sungusungu a success. Pimbwe social organization is relatively simple and includes no organization above the village level. The differences between Sukuma and Pimbwe are also reflected in how they play the Ultimatum Game. Sukuma offers in the Ultimatum Game are among the highest ever recorded. Mean offers made by Sukuma playing with an anonymous person from the same village were over 60 percent. When playing with someone from a different village, Sukuma players still offered, on average, more than half of the amount they had available. Mean offers made by Pimbwe players were lower in both types of game, and the contrast between within-village and different-village offers was much greater than among the Sukuma, perhaps reflecting the lack of institutions among the Pimbwe that involve more than one village. The contrasts between the Sukuma and Pimbwe may also be driving a process of cultural group selection as Pimbwe adopt Sukuma culture traits and, through intermarriage, become Sukuma themselves.[6]

Slave rebellions, and the lack thereof

••

Slavery has existed in a wide variety of societies throughout history, but the Atlantic slave trade was by far the largest. Between the fifteenth and nineteenth centuries, approximately ten million Africans were captured and shipped to the Americas, and millions of their descendants lived their entire lives as slaves.[7] Despite the fact that slaves outnumbered slave owners in many areas, slave rebellions did not occur often, and they were seldom successful. One of the rare exceptions occurred on board *La Amistad*, a slave ship, in 1839. After the ship docked in Connecticut, U.S. courts eventually found that they would-be slaves were free, and they returned to Africa.[8]

Few groups of people would seem to have a greater incentive to overcome the collective action dilemma than slaves. Realizing this, slaveholders went to great lengths to prevent slaves from assembling, becoming educated, and organizing. Those rebellions that did occur were dealt with very harshly, and, with important exceptions such as the Haitian rebellion, which not only freed thousands of slaves but also led to Haitian independence, they were almost never successful. The mutiny that occurred on board *La Amistad* in 1839 is an interesting exception.

Many factors combined to make the rebellion on *La Amistad* both possible and successful. *La Amistad* was not one of the infamous oceangoing slave ships of the Middle Passage, in which hundreds of captured Africans would be shackled below decks. Rather, it was a small schooner designed for the Cuban coastal slave trade. At the time of the mutiny, it was sailing from Havana to another Cuban port, carrying only fifty-three captives, almost all of them young adult men. Its crew was also small, consisting of a captain, his cabin boy, a cook, two sailors, and the two Spaniards who had bought the slaves and chartered the boat. The small number of captives made the collective action dilemma easier to solve, and the high ratio of captives to crew made the rebellion more likely to succeed.

A variety of conditions on board *La Amistad* also encouraged the rebellion and contributed to its success. Security was much more lax than on an oceangoing slave ship, making it possible for slaves not only to free themselves but also to find weapons, mostly in the form of large knives used to cut sugar cane. Food was in short supply, providing an added and immediate incentive for the rebellion. Because the captives were newly arrived from Africa and spoke no Spanish, communication between the captives and crew was difficult, and it was made worse by a joke that the cook told one of the captives. When asked, through hand signals, what the crew intended to do with the captives, he used gestures to indicate that they were to be slaughtered, their flesh salted and stored in barrels like beef, and eaten. The captive to whom the cook told his joke is known mostly by the English name he eventually adopted, Joseph Cinqué. Cinqué is the final piece of the *Amistad* puzzle. He appears to have been a good example of an entrepreneur, that is, someone who is willing to pay more than his share of the cost to help create a public good because he stands to gain so much from it. He also appears to have been an excellent leader and tactician who knew that he could count on help from a small number of other, like-minded captives.

Air traffic, coordinated and uncoordinated

Every day, tens of thousands of commercial aircraft take off and land around the world. Many airports are extraordinarily busy, with some handling thousands of flights a day. Despite all of that traffic in the air and on the ground, collisions between planes are very rare. One such rare and tragic event resulted in the worst air disaster in history. On March 27, 1977, on the Spanish island of Tenerife, a Boeing 747 owned by KLM collided on takeoff with a Pan Am 747 taxiing on the ground. All of the 248 people aboard the KLM plane died, as did 335 people on the Pan Am jet.

The disaster at Tenerife is a clear example of a failure to solve the problem of coordination. Many thousands of times a day at thousands of airports around the world, in contrast, the same problem is handled smoothly and routinely. At the world's many small airports that lack control towers, successful coordination of aircraft movements occurs thanks to clear rules and conventions regarding the movements of planes and effective communication among pilots. At larger airports, directions from airport control towers, usually aided by ground radar, play an important role. But on that day at Tenerife, several factors contributed to coordination failure, and disaster.

To begin with, neither plane was even supposed to be at Tenerife. Rather, they were scheduled to land on another of the Canary Islands, Gran Canaria, but all flights from that airport were diverted due to a terrorist bombing. This led to crowded conditions at Tenerife's relatively small airport, which had only one runway. The crowded conditions forced planes to taxi on the runway rather than on the adjacent taxiway, where other diverted aircraft were parked. To make matters worse, the airport was shrouded in heavy fog. Actions of the crews, particularly that of the KLM jet, also contributed to the failure to coordinate. Although the KLM jet could have refueled when it eventually reached its original destination of Gran Canaria, its pilot chose to refuel on Tenerife, apparently in an effort to save time. This added weight to the plane and hampered its ability to maneuver away from danger. The Pam Am crew failed to exit the runway at the taxiway indicated by ground control officials, perhaps because of confusion due to the fog and the absence of any signs to number the taxiways. Their confusion may also have led them to taxi more slowly than normal, which left them in the runway long enough for the disaster to occur. Finally, miscommunication, particularly between the KLM crew and the tower, seems to have led the pilot of the KLM plane to believe that he had been cleared for takeoff.

As we saw in chapter 5, coordination problems are solved through the creation of common knowledge about how to solve them and common knowledge that everyone concerned knows the solution (i.e., common metaknowledge). Clearly, both common knowledge and common meta-knowledge were missing at the Tenerife airport on March 27, 1977. The entire situation was unusual and unfamiliar to all concerned. Communications were faulty. Taxiways were not numbered. Perhaps most importantly, fog prevented the pilots and the tower from seeing the aircraft, and no ground radar existed to remedy that situation.[9]

Consilience, emergence, and the scientific division of labor

As we have seen, there are two broad types of problems that stymie co-operation: conflicts of interest, as in the collective action dilemma, and coordination problems. As the title of this book suggests, the problem that social and evolutionary scientists face when studying cooperation is a coordination problem. Because everyone wants the same thing—to better understand the diverse and complex phenomenon of human coopera-tion—there is no conflict of interests. The only thing missing for this co-ordination problem to be solved is the common knowledge necessary for scientists of all stripes to coordinate their efforts.

The social scientific and evolutionary biological approaches to behav-ior are, fundamentally, complementary to one another. Because they mostly deal with quite different phenomena, they have different vocabu-laries, methods, theories, and so on. Most of the time, that works well enough. For example, because plant biologists and political scientists rarely have much to say to each other, there isn't much call for them to develop a common scientific language. But when social and life scientists study the same phenomenon—such as cooperation—they need to find some common ground or risk simply talking past each other. By showing that cooperation is best understood when both approaches are brought to bear and by identifying ways in which this can be accomplished, we have tried to create a sort of Grand Central Terminal for this scientific coordination problem, a place where scientists of diverse backgrounds will have enough common knowledge that they can meet, shake hands, have intelligent conversations, and perhaps even agree with each other from time to time.

The idea that causal explanations generated by different scientific disciplines are complementary, not competitive, is not new. Borrowing a term from nineteenth-century philosopher of science William Whewell, Edward O. Wilson has referred to the "interlocking" of causal explanations offered by different disciplines as "consilience."[10] The idea here is simple: Because all phenomena in the universe are ultimately interconnected, so, too, should be our explanations of them. By avoiding independent and irreconcilable explanations for different phenomena, we gain a picture of reality that is more accurate than if different disciplines were to spin their tales (and, sometimes, chase their tails) in isolation from one another.

Exactly how different explanations interlock—and thus how different sciences are consilient—varies from case to case. One way arises from the simple fact that many phenomena are not produced by a single cause acting alone. Although scientists often make great progress by examining just one causal relationship at a time, many real-world phenomena are products of many different causes acting together. This is something with which we are all familiar from our daily lives: Did the automobile accident occur due to faulty brakes, one driver's cell phone use, the other driver's alcohol consumption, the rain on the highway, or some combination of factors? A full explanation might involve ideas from disciplines as diverse as physics, biology, and psychology. Another way different causal explanations can fit together is by focusing on factors that are at different causal distances from the phenomenon in question. As we explained earlier in this book, this is so routine among biologists that they have created a division of labor among themselves based on the causal distance in which they are most interested. Functional biologists focus on proximate causes, ontogenetic causes interest developmental biologists, ultimate or distal causes are adaptationists' main concern, and species' phylogenetic histories are paleontology's area of expertise. An explanation at one degree of causal distance is just as legitimate as an explanation from any other, but they must also all fit together coherently.

Though Wilson's case for consilience among the sciences is compelling, it is missing one element that we consider essential: emergence. By showing how different explanations can be relevant to different levels of emergent phenomena, emergence creates another route toward complementarity among explanations offered by different disciplines.[11] The social sciences study norms, conventions, institutions, patterns, and other phenomena that emerge from the interactions of evolved organisms. To fully understand why those organisms behave in ways that sometimes lead to emergence, we need to understand how they evolved. But an understanding of how those individuals evolved is not sufficient by itself to explain the broader social phenomena of interest to the social sciences. Those social phenomena have properties that require theories about them in their own right.[12] At the same time that evolutionary biological explanations of individuals' behavioral tendencies in no way undermine the social sciences, explanations of emergent social phenomena that make little reference to evolution do not necessarily violate the integrity or principles of the life sciences.

Although some see emergence as a barrier to explanation and reduction, we consider the identification of the roots of emergent phenomena in interactions among individual agents to be an important part of science's mission. This position is sometimes called "weak emergentism."[13] However, even weak emergentists like us acknowledge that there is more to study about emergent phenomena than simply how they emerged. Chemistry emerges from physics, but there is more to chemistry than physical chemistry. Chemists also study properties of compounds that simply have no meaning in terms of their constituent elements. Acidity, which has meaning only when applied to compounds, not elements, is a good example. Similarly, life emerges from chemistry, but there is much more to biology than biochemistry. Biologists also study things that have no real meaning at the chemical level: selection, niches, and trophic levels are just a few examples. While it may be possible to tell a causal tale at the level of physics and chemistry about any biological phenomenon, it will often not be a particularly interesting story. If there is more than one way that physics and chemistry can combine to produce a particular phenomenon of interest to biology, then such a move might also be misleading.[14] Signals, for example, can be produced through very different physical means (e.g., pheromones or bird songs), but because they have so much in common they are quite rightly the subject of a highly developed body of theory that does not depend on them having any particular physical basis. To borrow a phrase from philosopher Daniel Dennett, signals are "substrate neutral."[15]

The relationship between the study of individual organisms and social phenomena is analogous to the one between the physical sciences and biology. Although social phenomena emerge from interactions among individuals, there is more about them that is worthy of study than simply the characteristics of the individuals or the process of emergence. Reducing social phenomena to the actions of individuals may always be possible in principle, but often it may not be a very interesting or enlightening thing to do. And, as in biology, such a move might sometimes be misleading. For example, money, like signals, is substrate neutral. Although money may take many different physical forms (coins, paper currency, checks, etc.) and although dealing with it might involve a wide range of different actions by individuals (e.g., handing over bills versus swiping a credit card), all of the instances of money have more than enough in common to justify the existence of a great deal of theory about money that makes no assumptions about money's physical form or specific individual behaviors associated with it.

Exactly how different sciences interconnect may differ depending on which particular sciences we consider. Certainly there is a sense in which some sciences set the rules by which other sciences must play. In order for biology to become consilient with the rest of the natural sciences, for example, it had to accept the inviolability of the principles of the physical sciences. Contrary to the "vitalist" biologists of the nineteenth century, life could not contain some special kind of energy not found in nonliving things. Similarly, living things cannot violate the laws of thermodynamics or the law of gravity, and biological theories cannot propose that they do. However, this does not give physics and chemistry veto power over biological theories. If it did, evolutionary biology would have been stopped in its tracks in the nineteenth century when physicists' theories regarding the ages of the earth and sun did not give evolution much time in which to operate. Those theories turned out to be wrong, the earth turned out to be quite old, and evolutionary theory turned out to be right.[16] Consilience was achieved, but by changing theories in physics rather than in biology.

If the social sciences wish to become consilient with the rest of the sciences, they, too, must accept the fact that humans and the phenomena they produce cannot violate the laws of physics. Fortunately, no one (within the sciences, at least) has ever suggested otherwise. What *has* been suggested, time and again, is that social phenomena are somehow sui generis, forming a realm independent of all others and immune to explanations arising from such disciplines as biology and psychology. For example, such a position has often been quite standard in anthropology regarding the phenomenon of culture. *Omnia cultura ex cultura* (roughly, "all culture comes from culture") is the way that one prominent cultural anthropologist put it many years ago.[17] This view is as wrongheaded in the social sciences as vitalism was in biology. It is no more necessary for the social sciences to be cordoned off from biology in order to be legitimate in their own right than it was for biology to be cordoned off from chemistry or physics. Rather, biology's legitimacy as a science required that it become connected to the physical sciences. The same is true for the social sciences vis-à-vis biology and the physical sciences.

While there are no "laws of biology" or "laws of evolution" that are inviolable in the same way as the laws of physics, social scientists who propose that any particular human behavior, culture, or social arrangement is independent of or contrary to what we think we know about human evolution have an obligation to explain how such a thing could have come about. This is not an impossible task. Indeed, there is a small but important specialization within anthropology devoted to precisely that endeavor, that is, identifying situations in which culture and genes lead behavior in different directions. Research in this area has shed light on such phenomena as food taboos, cannibalism, headhunting, and religion.[18] The picture emerging from this work is that culture can indeed lead people to do things that run contrary to the expectations of evolutionary biology, but deviations that are voluntary and that lead to sharp reductions in health and reproductive success for more than a few people are usually rather short-lived.

Some of the social sciences' resistance to consilience may be a result of the misleading dichotomy between "natural" and "artificial." Friedrich Hayek, who traced this intellectual tradition back to the ancient Greeks, blamed it for some of the difficulty subsequent scholars have had in appreciating emergent social phenomena.[19] If everything is either "natural" and so involving no human action or "artificial" and so the product of deliberate human action, how do we account for phenomena that are, as Adam Ferguson put it, "the result of human action but not of human design"? This same dichotomy is behind the idea that the "natural sciences" are distinct from the "social sciences." Certainly some of what social scientists study is not "natural" in the sense that it is deliberately planned and created by people—constitutional government, for example. But social scientists also study many things that are not deliberately planned or even foreseen by the people whose actions create them. As Hayek put it, "*Culture is neither natural nor artificial, neither genetically transmitted nor rationally designed*. It is a tradition of learnt rules of conduct which have never been 'invented' and whose functions the acting individuals usually do not understand."[20] Forcing such phenomena into one category or the other inevitably obscures something interesting and vital about them. Better to eschew both the false dichotomy between "natural" and "artificial" and the one between the natural and social sciences. Doing so will lead to better scientific understanding of cooperation and many other phenomena.

Another barrier to consilience may be some common readings—or misreadings—of classic texts. Consider, for example, Emile Durkheim's famous dictum about the distinctiveness of psychological and social phenomena: "every time that a social phenomenon is directly explained by a psychological phenomenon, we can be sure that the explanation is false." Readers might be surprised to learn that we think Durkheim was on the right track. The key to understanding Durkheim's real point is to read beyond that oft-quoted sentence. It then becomes clear that he understood the concept of emergence. Here is the sentence that immediately precedes the one above: "In a word, there is between psychology and sociology the same break in continuity as between biology and the physicochemical sciences."[21] Durkheim understood that because life was an emergent property of physics and chemistry there needed to be a life science separate from the physical sciences. Similarly, he argued that because social phenomena emerge from interactions among individuals, we need a science that deals with emergent social phenomena in their own right. In his words, "society is not a mere sum of individuals. Rather, the system formed by their association represents a specific reality which has its own characteristics."[22]

Making the collaboration work

While we agree with Durkheim that the complementarity of the physical and life sciences matches that of the psychological and social sciences, it is not enough simply for evolutionary and social scientists to accept this idea and move on. A more fundamental reform is necessary. Just as the physical sciences undergird the life sciences, all social theory is underlain by some attempt to understand individual human psychology. But while the life sciences can call upon the well-established and understood principles of modern physics and chemistry, the psychological notions that underlie most current social science are tacit rather than explicit and often consist mostly of intuition, assumptions, folklore, generalizations, and educated guesswork. Take, for example, social scientific work on the family. The large scholarly literature on "dysfunctional families" takes for granted that families should be "functional," that is, nurturing, loving, caring, and supportive. But why should they be like that? This may seem like the norm or ideal among members of our species, but many other species show few if any of these characteristics, and many do not even form social units that anyone would call a "family." Unless we place humans in this kind of comparative context, our assumptions about what constitutes a "functional" family will remain unexamined.[23]

To make matters worse, each of the social sciences has its own pet theory of individual psychology, and most of them are not particularly well developed. For example, cultural anthropologists have long argued that culture shapes the mind and behavior in powerful ways. In some instances, they certainly must be right, but in others, culture seems to influence thought and behavior weakly or not at all.[24] With few exceptions, anthropologists have treated the idea that culture shapes behavior as an unquestionable truism rather than as a hypothesis to be tested. As a result, our understanding of the actual causal connections among culture, the mind, and behavior is quite poor. The evolutionary approach suggests a way forward: Consider how selection pressures may have shaped us to be easily influenced by some types of culture traits and resistant to others. As suggested by cultural transmission theorists, we may have a bias in favor of traits displayed by people who are admirable or successful in some way, even if the traits in question play no role in their success.[25] As suggested in chapter 6, another possibility is that, given that coordination can take place only if behavior is allowed to follow the dictates of social coordination norms, we may have evolved to be choosier about culture traits that concern individual behaviors than about those that concern behaviors that must be coordinated with others.

Or, consider Olson's dismissal of the idea that solidary benefits are necessary to explain why people join groups. As we have seen, a great deal of subsequent scholarship has shown that people *do* join groups because they enjoy the feeling of solidarity they experience when working with others to further common values. The civil rights movement (which, ironically, had some of its greatest achievements while Olson was writing his book) described in chapter 3 is a case in point.[26] Olson's position reflected the simple theory of human psychology underlying his approach to human behavior, one that he shared with many other economists and political scientists, which equated rational self-interest in general with rational self-interest in pursuit of material, tangible benefits. But because selection acts not on an organism's ability to acquire material things but rather on its ability to reproduce, our thinking about motivations should include not only material ones but also anything that, if acquired, might have enhanced our ancestors' reproductive rates. As we saw in chapters 2 and 4, an evolutionary account of why people are motivated by solidary benefits is already at hand. Thanks to natural selection, people seek out coalitions and enjoy coordinating their behaviors with those of others. In addition to the motivations, natural selection has also provided people with the cognitive tools, such as Theory of Mind, necessary to realize those desires.

In chapter 3, we described a finding by political scientist Terry Moe that might be particularly ripe for an evolutionary explanation: people tend to overestimate the positive impact they have on the groups they join, leading them to join groups when a fully informed rational decision maker would not do so.[27] One possible evolutionary explanation of this finding begins with the simple observation that most people are not particularly good at understanding large numbers. Why would they be? Although the modern world may force us to deal with large numbers every day, for our ancestors, who lived in small groups and had no money, small numbers were the order of the day. Even today, many languages have counting systems that amount to nothing more than "one," "two," and "many." Even something as commonplace and essential to today's society as voting may rely upon the difficulty we have with large numbers. Although it is a sacrilege in a democratic society to say that a single vote does not count, the reality is that most of us will live our entire lives dutifully voting in election after election without ever making a difference in who gets elected. The numbers involved are simply too large for one vote to matter for very much in most elections. Yet people do vote, and a tendency to overestimate one vote's impact may be a reason why. Other reasons to vote, such as the opportunity to be seen by one's neighbors as doing one's civic duty, certainly also exist, although Oregon's experience with mail-in ballots, which increased rather than decreased voter "turnout," suggests that something else is going on.[28] Thus, even something as important to our way of life as democracy may ultimately rest upon a mismatch between our current environment and the one in which we currently live. This is what we described in chapter 2 as a "novel environment" argument: because adaptations arise in specific environments, novel environments may produce unexpected or even maladaptive responses. To give this a more positive spin, we could consider democracy's reliance upon mass innumeracy one of the many clever work-arounds that, according to Peter J. Richerson and Robert Boyd, typify modern human social arrangements.[29]

In chapter 2, we briefly described another idea from evolutionary psychology that may help explain Moe's finding. Ideally, natural selection would have designed our minds with the ability always to make the right decision, accurately weighing the costs and benefits of our different options. In reality, we make errors, and those errors come with costs. If the cost of making one kind of error is much larger than that of making another kind, selection pressure on how we make that kind of decision will be asymmetrical. A tendency to make more of one kind of relatively low-cost kind of error rather than more of a relatively high-cost kind may be a design feature, not a flaw, of the human mind. This is the idea behind error management theory, as developed by Martie Haselton and her colleagues, and the smoke-detector principle, as developed by Randy Nesse.[30] Applying this idea to Moe's observation, it may be that the error of contributing to a public good and having that contribution not bear fruit is often a small price to pay compared to the error of failing to help create a public good from which one would have benefitted greatly. Given that our ancestors lived in small groups and among kin, easing the collective action dilemma and creating an indirect fitness benefit, this could easily have pushed our psychology in the direction of erring on the side of participation by overestimating the degree to which our contributions really matter to the success of the collective action. These ideas, while speculative, are testable. Our point is not that the evolutionary approach has all of the answers. Rather, we believe that it asks the right kind of questions and offers the right kinds of insights to provide social scientists with a much improved understanding of the kind of organism they are trying to understand.

A chapter-by-chapter summary

In chapter 1, we argued that a phenomenon as diverse as cooperation requires the eclectic approach of the historical sciences rather than the more purely deductive approach of the theoretical or Newtonian sciences.[31] At about the same time in the mid-1960s, Mancur Olson and George Williams wrote influential books that addressed quite similar sorts of questions: how to study social phenomena (Olson) and adaptations (Williams) without the assumption that people and other organisms simply are driven to do what is best for the group? To pull together research from these different traditions, one needs to appreciate the fact that different explanations of the same phenomenon can be complementary rather than conflicting. Although many evolutionary theorists equate cooperation and altruism, cooperation is best defined as it always has been: individuals working together. Because altruism is involved in some but not all cases of cooperation, the two concepts should remain distinct. Two types of problems may prevent cooperation: conflicts of interest and lack of common knowledge. Conflicts of interest are the key to the collective action dilemma, a situation in which everyone who might benefit from cooperation is also tempted by the prospect of free riding on the efforts of others. Even where there are no conflicts of interest, a lack of common knowledge can also prevent cooperation, a phenomenon known as a coordination problem.

Chapter 2 introduced the concept of adaptation. Evolutionary biologist George C. Williams argued that adaptation "is a special and onerous concept that should be used only where it is really necessary." The study of adaptations focuses on how selection—natural, sexual, social, and artificial—designs organisms to solve the problems presented to them by the environments in which they live. The study of adaptation complements other causal explanations in biology, such as those that emphasize proximate mechanisms, those that focus on the development of the organisms, and those that focus on the organism's evolutionary heritage. Each adaptation reflects both the phylogenetic history of the organism and the specific environment in which it arose. When organisms experience evolutionarily novel environments, their adaptations to past environments may no longer reliably produce adaptive outcomes. Some apparently maladap-

184 • Chapter 8

tive outcomes, however, may actually be design features, rather than flaws. Specifically, organisms may be designed by selection to accept high rates of cheap errors in exchange for avoiding expensive ones.

Culture, defined as socially transmitted information, is phylogenetically widespread, but its greatest development has clearly been among humans. As an important part of the environment in which humans live, culture has the power to influence which genes reproduce. Such gene-culture coevolution may be responsible for some of the adaptations that help humans cooperate, such as our ability to identify people who violate social rules. Language is one aspect of culture that may have been particularly important to helping humans achieve cooperation in degrees not seen among our nonhuman relatives.

When writing *Adaptation and Natural Selection*, one of Williams's main purposes was to present a critique of the idea that selection works through the differential survival and reproduction of groups. Although most biologists follow Williams in relegating group selection to a minor role in explaining the behavior of most organisms, group selection does have its advocates. Some of the efforts to revive group selection have relied upon a definition of "group" that is so broad as to make it virtually meaningless. What role, if any, that biological group selection has played in the evolution of adaptations that underlie cooperation in humans is still hotly debated. As with all scientific debates, the only way to settle the matter will be through tests of competing hypotheses.

Many social scientists are concerned that explanations of behaviors, particularly unpleasant ones, in terms of natural selection will lead to moral justifications of those behaviors. Such a leap from what is "natural" to what is "good" is known among evolutionary scholars as the naturalistic fallacy, and it is judiciously avoided. Knowing that something is a product of selection provides no information about its moral standing.

In chapter 3, we reviewed the social science literature on cooperation, emphasizing work that has been done since the first edition of Olson's *The Logic of Collective Action*. Olson's model showed mathematically why under most circumstances we should not expect an individual to contribute to a collective good. Of course, we know from observing the world that groups do in fact form and collective goods are in fact provided, and Olson and those who followed have suggested and tested hypotheses about the conditions under which this is likely to happen. The key to the provision of most collective goods are selective benefits—material, solidary, and purposive or expressive—that serve as private side payments to encourage the individual to join the collective undertaking. Entrepreneurs and patrons help provide these selective benefits. These theories all involve changing the structure of the collective situation in such a way that the individual ends up better off by participating in the collective effort rather than sitting it out. One limitation of these theories is that while they suggest that individuals do what is "best" for them or that provides them the greatest benefits, they do not have any very satisfying answer to what it is that people want. Olson and the economists assumed what people wanted was material benefits. Political scientists and sociologists have been less than happy with that answer but still lack any convincing and theoretically grounded alternative.

Because every approach in the social sciences begins with some theory of the individual, chapter 4 explored the ways in which our evolved psychology contributes to our ability to cooperate. Research on genes, brain function, and hormones suggests that many of these abilities have biological, heritable underpinnings and so are likely to be products of selection. The benefits of reciprocity would have favored individuals who are willing to engage in mutually beneficial reciprocal transactions, but the risk of being cheated would have led to selection pressures in favor of an additional ability: identifying good cooperative partners and avoiding cheaters. Even infants seem to have this ability, choosing to play with toys that have been depicted as helpful to other toys over toys that have been depicted as harmful. As a result, selection may also have favored those who went out of their way to display their generosity and cooperativeness and who worried about their reputations.

Chapter 5 reviewed the role that organizations play in fostering cooperation. Many successful organizations simply build upon the successes of past organizations, using their ability to mobilize and motivate people and to frame issues to overcome the collective action dilemma. The literature on common-pool resource management provides many good examples of both successful and unsuccessful organizations. Some of the keys to successful common-pool resource management include clearly defined boundaries, rules that are appropriate to local conditions, some way of changing the rules, some way of monitoring members' compliance with the rules, a system of graduated sanctions for rule violators, and a system for conflict resolution. Once organizations and other kinds of groups exist, they can undergo a process of selection that is analogous to but quite different from selection among biologically defined groups. This cultural group selection may help explain our remarkably flexible coalitional psychology, that is, our ability to form heartfelt bonds with the groups with which we cooperate but still shift those feelings to other groups as better opportunities arise.

In collective action problems, cooperation is stymied by conflicts of interest. In coordination problems, the main roadblock to cooperation is a lack of common information: everyone would benefit if coordination were to be achieved, but they may not know how to do it. As we explained in chapter 6, the key to solving coordination problems is common knowledge. In addition, people must have common metaknowledge, that is, common knowledge that there is common knowledge. Common knowledge alone is insufficient because without common metaknowledge no one will be confident that others will arrive at the same solution to the coordination problem. Our ability to solve coordination problems may reflect how important such solutions were to our ancestors, and aspects of our physiology, psychology, and culture may reflect this evolutionary heritage. Unlike other primates, our irises are set off in fields of white, which makes it easier for others to see what we are looking at. Evolutionary anthropologists have suggested that this may have evolved because it facilities shared intentionality. Humans are also very good at reading each other's minds, an ability known as Theory of Mind. Such an ability is crucial to social coordination. Language is a particularly useful tool for social coordination, and language itself is essentially a convention that allows people to share information and coordinate their behaviors. Ethnic markers may also facilitate coordination by making it easy for people to identify others with whom they are likely to share social coordination norms.

Many kinds of coordination problems exist. In some, everyone agrees on the best outcome, and coordination is the only problem. In others, coordination problems are combined with conflicts of interest. Although coordination problems can be solved, they do not necessarily have a best or optimal solution. Any solution may be better than none, and many different solutions may be possible. Which one actually does the job is often more a matter of history and culture than rational calculation. This makes coordination problems more amenable to ethnographic and historic approaches than deductive, mathematical ones. Good ethnographic examples of social coordination norms come from research on whale hunters in Indonesia, rice farmers in Bali, herders in East Africa, and social movements in the United States. Some of this work suggests that solving coordination problems may be so valuable to people that we are more susceptible to social coordination norms than to culture traits that do not concern social coordination.

Solutions to coordination problems can arise in many ways. In chapter 7, we explored one of the more interesting possibilities: emergence. In some circumstances, solutions to coordination problems can emerge spontaneously simply through the interactions of individuals. Each encounter between individuals taps into a network of previous encounters with other individuals, allowing information to be pooled and a consensus to be formed. The idea of emergence has a long history in the social sciences. During the Scottish Enlightenment, social emergence was explored by such figures as David Hume, Adam Ferguson, and Adam Smith. Smith's famous phrase "the invisible hand," though widely misunderstood, captures the logic of social emergence: though something may appear to be the product of deliberate planning, it actually emerges with no planning at all. Darwin's theory of natural selection is a good example of an invisible-hand explanation because it showed that although organisms appear to have been deliberately designed, they were not. Another classic example is Carl Menger's explanation of how money may have emerged spontaneously through interactions in a barter economy.

Not everything that emerges spontaneously is necessarily beneficial. Traffic jams and racially segregated housing may also emerge and persist through invisible-hand processes. Such processes can be modeled both verbally and mathematically, but agent-based models are proving very useful tools in the study of social emergence. In some cases, social emergence takes advantage of mathematical properties. For example, many phenomena follow a pattern, known as a power law curve, in which some varieties are extremely common and others quite rare. This pattern can emerge spontaneously simply through random copying processes as common varieties become more common and rare varieties become more rare. Power law curves help explain a wide variety of phenomena including how strikes spread among workers and how bandwagons form around particular items on the political agenda.

The evolutionary and social scientific approaches to behavior, including cooperation, must and will eventually fit together. This is true for a very simple reason: the universe itself fits together. To make this possible, scholars in both areas need an appreciation of the complementarity of their various approaches, which was the focus of our final chapter. "Consilience" is a convenient label for this concept. In addition, they need an appreciation of emergence. Because living things have emergent properties that cannot be explained simply in terms of physics and chemistry, we need a science of biology. Similarly, social phenomena have emergent properties that require their own theories, not simply those of biology or psychology.

This book is about cooperation, in not one but two ways. First, it is about cooperation as a social phenomenon. Why do people cooperate sometimes but not others? Why do people cooperate more than their nonhuman relatives? Second, this book is about cooperation among the sciences. If we are to fully understand not only cooperation but also everything else about the human experience, we are going to need ideas from many fields, but in particular from the social and evolutionary sciences. By explaining terms, concepts, and theories emerging from both of these areas, we hope to have created a space in which all students of cooperation can meet and exchange ideas.

Notes

1. Olson 1965:2.
2. Renz et al. 2002.
3. For example, see Baumgartner and Leech (1998).
4. For example, see Cronk (1999).

CHAPTER 1
COOPERATION, COORDINATION, AND COLLECTIVE ACTION

1. Hirt et al. 2008.
2. Ostrom 1990.
3. Paciotti et al. 2005; Paciotti and Borgerhoff Mulder 2004; Paciotti and Hadley 2003, 2004.
4. Curtin 1972.
5. Jones 1987.
6. This description is based mainly on a Wikipedia page about the disaster (http://en.wikipedia.org/wiki/Tenerife_airport_disaster).
7. Hayek 1955; Aberle 1987; Simpson 1980; Kitts 1981.
8. Kitts 1981.
9. Hayek 1955.
10. Kitts 1981.
11. Olson 1965; Williams 1966.
12. Olson 1965:2, emphasis removed.
13. Salisbury 1969.
14. The term "replicator" is from Dawkins (1976), who later elaborated on the idea (Dawkins 1983). We changed the term slightly in order to distinguish things that make copies of themselves from things that make copies of other things (e.g., photocopiers).
15. Wilson 1998.
16. Binmore 2007; Gintis 2009.
17. Hardin 1992; Ainsworth 2002; Oliver 1993; Oliver and Myers 2002; Barash 2003.
18. Not to be confused with Cupertino (see http://en.wikipedia.org/wiki/Cupertino_effect).
19. Important exceptions do exist. See, for example, Alvard and Nolin (2002), Dugatkin (1997), Clements and Stephens (1995), and Smith (2003).
20. Nowak 2006:90.
21. Henrich and Henrich 2007:37.
22. Bowles and Gintis 2003:429–30.

23. Lehmann et al. 2008:664.

24. Evolutionary biologist Andrew F. G. Bourke (2011) recently suggested a route around this terminological problem. He used the phrase "narrow sense cooperation" to refer, more or less, to what we are calling cooperation and the phrase "broad sense cooperation" to refer to cooperation plus altruism. If we simply use cooperation to mean "working together" and altruism to refer to self-sacrificial behaviors, then such jargon gymnastics are unnecessary.

25. Trivers 1971.

26. Hamilton 1996; West et al. 2011.

27. See Hamilton (1964) and Dawkins (1976).

28. Richerson et al. 2003.

29. Lloyd 1833:18.

30. Dawes 1980; Kollock 1998.

31. Salisbury 1969.

32. See Ullmann-Margalit (1977) for more on the distinctions among coordination, cooperation, and collective action.

33. For example, Bowles and Gintis (2003).

34. Schelling 1960.

35. Williams 1966:4.

36. Chwe 2001.

CHAPTER 2
ADAPTATION

1. Williams 1966:4.

2. Mayr 1961; Tinbergen 1963. Laland et al. (2011) make the important point that causation among the various levels should not be thought to flow in only one direction; reciprocal causation not only is possible but might be quite common.

3. West et al. 2011.

4. de Quervain et al. 2004:1254.

5. Williams 1966:9.

6. West-Eberhard 1975; Brown 1983; Clements and Stephens 1995; see also Dugatkin (1997).

7. Connor 1986.

8. Blurton Jones 1984.

9. Boone 1998.

10. Alvard 2001.

11. Sugden 1986; Doebeli and Hauert 2005.

12. Nesse 2007; West-Eberhard 1979; Boehm 2008.

13. These adaptations and their impacts on human cooperation will be examined more fully in later chapters.

14. For a review of sexual selection among humans, see Low (2001).

15. Williams 1966:264.

16. Linden 2007:246, emphasis in original.

17. For humans, see Brown et al. (2005). For zebra finches, see Swaddle and Cuthill (1994).

18. Quinlan and Quinlan 2008.
19. Gould and Vrba 1982.
20. Dennett 1995.
21. For more on this idea, see Darwin (1859:435).
22. Buss et al. 1998.
23. Richerson and Boyd 1999.
24. DeBruine 2002.
25. Oates and Wilson 2002.
26. On kin term manipulation, see Johnson (1987), Salmon (1988), and Chagnon (1988, 2000). On the use of fictive kin terms by religions, see Qirko (2001, 2002, 2004a) and Soler (2008, in press).
27. Qirko 2004b, 2009.
28. Ambrose 1992.
29. Bowlby 1969.
30. See, for example, Barkow et al. (1992).
31. Irons 1998.
32. Harcourt-Smith and Aiello 2004.
33. Perry et al. 2007; Laland et al. 2010.
34. See, for example, Newton-Fisher (2006).
35. Flinn and Alexander 2007; Flinn and Coe 2007.
36. For more on niche construction, see Odling-Smee et al. (2003).
37. Even interest group entrepreneurs create niches for their groups. See Gray and Lowery (1996).
38. This insight comes from Drew Gerkey, who has studied cooperation among salmon fishers and reindeer herders in Kamchatka (Gerkey 2010).
39. Haselton and Buss 2000; Haselton and Nettle 2006; Nesse 2001, 2005.
40. Atran and Norenzayan 2004; Guthrie 2001; Shermer 1998; Foster and Kokko 2009.
41. Johnson 2009.
42. Tylor 1871:1.
43. Cronk 1995, 1999; Alvard 2003a; Durham 1991.
44. The cake example is derived from Geertz (1973:250).
45. Keesing 1974.
46. See, for example, Boyd and Richerson (1985), Cavalli-Sforza and Feldman (1981), Durham (1991).
47. Nozick 1974; Cronk 1999.
48. Laland and Williams 1998; Dugatkin and Godin 1993.
49. Terkel 1996.
50. Rendell and Whitehead 2001; Marler and Tamura 1964.
51. Rendell and Whitehead 2001.
52. McGrew 1992; Perry et al. 2003.
53. See, for example, Kendal et al. (2009).
54. Durham 1992.
55. Durham 1992; Tishkoff et al. 2007.
56. Livingstone 1958.
57. Laland and Williams 1998; Dugatkin and Godin 1993.
58. Reader et al. 2008; Waynforth 2007; Yorzinski and Platt 2010; Jones et al. 2007.

59. Eric Alden Smith (2003, 2010) has made this argument forcefully and convincingly.

60. Brannon 2005; Frank et al. 2008; see also Gelman and Gallistel (2004).

61. Frank et al. 2008:823.

62. Color: Winawer et al. 2007; navigation: Hermer-Vazquez et al. 1999; perception of emotion: Barrett et al. 2007.

63. Dawkins 1976; Burt and Trivers 2008.

64. Rawls 1971; Ridley 2001.

65. Emerson 1960; Williams 1966:225.

66. See Carr-Saunders (1922). Among the many problems with Emerson's theory is the fact that dying of old age is actually an unusual event among wild species (Williams 1966:226). Carr-Saunders's logic turns out to be unsound, as well. Although women in hunting and gathering societies do exert control over their reproductive rates, they do so in ways that maximize rather than minimize the number of their children who survive to adulthood, a pattern consistent with individual-level selection, not group selection (Blurton Jones 1987; see also Lack 1954).

67. Wynne-Edwards 1962.

68. Williams 1966:218.

69. Williams 1966:4–5.

70. Williams 1966:5.

71. Dawkins 1976.

72. Lewontin and Dunn 1960.

73. Avilés 1993.

74. Williams 1966:8.

75. See Maynard Smith (1964).

76. E.g., Trivers (1985).

77. Maynard Smith 1976:277.

78. Bernhard et al. 2006:912, emphasis added.

79. Price 1970.

80. For a good, readable explanation of the Price equation and many other models important in evolutionary biology, see McElreath and Boyd (2007).

81. See Okasha (2006) for a recent analysis of multilevel selection theory and Okasha (2010) for a brief update. In addition to Okasha's book, readers interested in the group selectionist debate would do well to start with several recent articles by Stuart West, Alan Grafen, Andy Gardner, and their colleagues (Gardner 2009; Gardner and Grafen 2009; West et al. 2007; West et al. 2010). In 2009, the London Evolutionary Research Network (LERN) sponsored a debate on group selection featuring talks by Okasha and West, along with biologist Mark Pagel and economist Herbert Gintis. A link to videos of those talks can be found on LERN's website (londonevolution.net). David Sloan Wilson has explained his current thinking on these issues in his "Evolution for Everyone" blog (http://scienceblogs.com/evolution/).

82. Sober and Wilson 1998. See also Wilson and Sober (1994).

83. Hamilton 1975.

84. Hamilton 1975:134.

85. Williams 1966:93.

86. Maynard Smith 1976, 1998.
87. Okasha 2006:184.
88. Maynard Smith 1998:639–40.
89. Wilson and Sober 1994.
90. Mange 1964.
91. Chagnon 1975.
92. Hurd 1983.
93. Rhesus macaques, for example: Chepko-Sade and Olivier (1979).
94. Cronk 1994.
95. Gintis 2000:169.
96. Baumard 2011; Burnham and Johnson 2005; Guala 2011; Johnson et al. 2008; Price 2008; Rankin 2011; Trivers 2004, 2006.
97. Sahlins 1965.
98. Levitt and List 2007.
99. Cherry et al. 2002.
100. See also Hoffman et al. (1994).
101. Bardsley 2008
102. Lehmann et al. 2008.
103. Kümmerli et al. 2010.
104. Marlowe et al. 2008.
105. Nikiforakis 2007.
106. Burnham and Johnson 2005.
107. Locke 2005.
108. Matos and Schlupp 2005.
109. For example, Fehr and Henrich (2003).
110. Trivers (2006) uses the experience of watching a horror movie rather than riding a roller coaster to make this point. The terror one experiences while watching a movie may be a result of one's feelings of empathy for the characters, and we can imagine someone arguing that our ability to have such feelings may be a product of group selection. We have chosen the roller coast to make this point because one's fears while plunging toward the ground at high speed are likely to be more focused on one's own safety rather than on that of those who happen to be sharing the experience.
111. Frank 1988.
112. Williams 1966:233.
113. Kurzban and Aktipis 2007.
114. Moore 1903.
115. See, for example, Shweder (2005) on female genital mutilation.

CHAPTER 3
THE LOGIC OF *LOGIC*, AND BEYOND

1. The birth rate is from the Population Connection's website (http://www.populationconnection.org/site/DocServer/2011_Dem_Facts.pdf?docID=2362). The death rate is from the CIA's World Factbook (https://www.cia.gov/library/publications/the-world-factbook/geos/xx.html). The CIA's estimate of the birth

rate is slightly lower than that of the Population Connection, but not by much (134 million per year).

2. http://www.populationconnection.org/site/PageServer?pagename =about_us.

3. Tillock and Morrison 1979.

4. Tillock and Morrison 1979:149.

5. Olson 1965:1.

6. Olson 1965:17.

7. Parsons and Bales 1955:9.

8. Truman 1951:57.

9. Olson 1965:16.

10. Olson 1965:2, emphasis in original.

11. Olson 1965:27–28.

12. Of course, the creation of the state in the first place required that the collective action dilemma be overcome, but, thankfully, that is beyond the scope of this discussion. We recommend Carneiro (1970) and Fukuyama (2011).

13. This idea comes from Clark and Wilson (1961); we will discuss it further in the next section.

14. AARP is now the organization's official name. It was previously known as the American Association of Retired Persons.

15. http://www.aarp.org/benefits-discounts/.

16. Walker 1991; Baumgartner and Leech 2001.

17. Walker 1991:31.

18. For reviews of this literature, see Baumgartner and Leech (1998), Hardin (1982), McAdam et al. (1996), or Ostrom (1998).

19. Clark and Wilson 1961.

20. Salisbury 1969. Salisbury preferred to call these benefits "expressive" benefits to avoid the implication that it was necessary to achieve the collective good before the selective benefits could be attained. "Expressive" goods come from speaking or acting, whether or not the collective effort is successful.

21. Olson 1965:61.

22. Levy and Chavez 1975:xxviii.

23. Lichbach 1994, 1995.

24. Moe 1980a, 1981.

25. See, e.g., Hardin 1982; Frohlich and Oppenheimer 1970.

26. Chong 1991:60; Fager 1974.

27. Chong 1991:61.

28. Although senior citizens today are among the best organized citizens in the United States and the AARP is the largest citizen membership group in the country, political scientist Andrea Campbell (2003, 2005) has shown that mobilization did not occur until the 1960s, well *after* the passage of Social Security and the Older Americans Act of 1965.

29. Amenta 2006:40.

30. Amenta 2006:58.

31. This is the main reason why many of Olson's critics have pointed out that Olson's distinction between large and small groups does not matter mathematically to his logic (see Hardin 1982; Frohlich and Oppenheimer 1970). Although

empirically "large" groups are more likely to face efficacy problems—as well as problems of monitoring shirking and rewarding participation—it is the lack of efficacy and lack of control that are theoretically important. Even large groups can succeed if they overcome the problems of efficacy and monitoring. Connecting the efficacy problem to the entrepreneurial solution, Oliver (1984) found that people who step forward as community leaders are more highly educated than their fellows but are not necessarily more optimistic about the probability that their neighbors will mobilize. In fact, they agree to serve as community leaders in part because they believe if they do not do so, no one else will. They feel efficacious compared to their fellow group members.

32. Walker 1983.
33. United Nations 2008; Simons 2012.
34. Rudé 1981.
35. Hardin 1982:42.
36. Rothenberg 1988, 1992.
37. Knoke 1988, 1990.
38. Muller and Opp 1986, 1987.
39. Marwell and Ames 1979, 1980, 1981.
40. Marwell and Ames 1979.
41. For a review of this literature, see Ostrom 2003.
42. Lichbach 1994:417.
43. King and Walker 1992:397.
44. It is beyond the scope of this book to discuss the advantages and disadvantages of this approach to public goods, which is the public interest group equivalent of what economists call rent seeking. See Buchanan et al. 1980; Leech 2006; Parker 1996; Tullock 1967.
45. For reviews of this literature, see Ledyard 1995; Sally 1995.
46. Our goal here is not to weigh in on one side or another in the arguments within political science over the advantages and disadvantages of "rational choice" theory. The question here is not so much whether the decisions are rational or irrational in a strict logical sense, but whether they are *sensible* for the individual human as an evolved human. Perhaps we should call it "sensible choice" theory. We expect decisions by humans—no matter what senses or methods are used to reach those decisions—to be goal driven and appropriate for those goals insofar as they are able, as evolved humans, to act and to choose appropriately. Research into whether these choices are cognitive, psychological, or emotional is fascinating, but should not detract from the central hypothesis that all forms of decision making should be expected to work in the individual's best interest in the long run. See chapter 2 for a discussion of what that long-run best interest is and where it comes from. For other discussions that take a broad view of rationality, see Chong (2000), Frank (1988), and Jasper (1998, 2011) on why emotional responses are not necessarily "irrational."
47. Moe (1980b) provides these figures broken down by group. We calculated these percentages by summing Moe's findings from across the five groups.
48. Although if Moe were to compare members to nonmembers, we would expect overestimation of efficacy to be more common among the members.

CHAPTER 4
COOPERATION AND THE INDIVIDUAL

1.   See Alford and Hibbing 2008 for a review of these studies and an excellent discussion of the broader role of biology in political science. Also see Alford et al. 2005; Fowler et al. 2008.

2.   Fowler and Dawes 2008.

3.   Decety et al. 2004; Haselhuhn and Mellers 2005; McCabe et al. 2001; Rilling et al. 2007; Rilling et al. 2008; Tabibnia and Lieberman 2007; Yoshida et al. 2010.

4.   Tabibnia and Lieberman 2007.

5.   Kosfeld et al 2005:673.

6.   De Dreu et al. 2010; De Dreu et al. 2011.

7.   Eisenegger et al. 2010.

8.   Sutton 2010:1.

9.   Binmore 2007:76.

10. Hume 1740: pt. 2, sec. 5.

11. Mauss [1922] 1990; Sahlins 1965.

12. See Aumann (1981).

13. Trivers 1971.

14. Axelrod 1984.

15. Aumann 1981:413.

16. Axelrod 1984; see also Axelrod and Hamilton 1981.

17. Nowak 2006.

18. Aktipis 2004, 2011.

19. West et al. 2011; see also Henrich and Boyd (2001).

20. Dugatkin and Sih 1998; Brown and Moore 2000.

21. Buss et al. 1990.

22. Campbell et al. 2009.

23. Carre and McCormick 2008; Roney et al. 2006.

24. Hamlin et al. 2007.

25. Mealey et al. 1996; Oda 1997.

26. Yamagishi et al. 2003.

27. Brown et al. 2003; Oda et al. 2009a; Oda et al. 2009b.

28. Sell et al. 2009.

29. Price 2006; see also Humphrey (1997). Price framed his findings in terms of a controversial approach to cooperation called "green beard theory." The idea behind green beard theory is simple: If cooperative individuals could simply identify each other, it would be easy for cooperative behavior to spread in a population. As W. D. Hamilton recognized, one way to do this would be for there to be some outward marker of one's willingness to cooperate with others who have the same marker and the same willingness to cooperate with others who have it. Ever since Richard Dawkins made the somewhat facetious suggestion that green beards might do the trick, this has been known as green beard theory (Dawkins 1976). For this to work as the basis for a genetically encoded tendency for cooperativeness, then the genes that code for the cooperative behavior, those that code for the green beard, and those that code for the ability to recognize green beards

must all be closely linked. As Dawkins pointed out, the problem is that such linkage is unlikely to last very long. When the gene for cooperativeness is inherited separately from the gene for the green beard, green beards will no longer be reliable indicators of a willingness to cooperate with others who have green beards. When that happens, uncooperative individuals with green beards will abound, refusing to cooperate with other green-bearded individuals but benefitting from the generosity of individuals for whom green beards remain linked with cooperativeness. Furthermore, because green beard genes are indicators of the presence of only a single gene (or a bundle of tightly linked genes) rather than genetic similarity between individuals across their entire genomes, selection on the rest of the genome will favor genes that lead individuals to ignore green beards (West et al. 2007; Gardner and West 2010). Despite how unlikely it seems at first glance, green beard theory continues to receive attention from theorists (e.g., Riolo et al. 2001, 2002; Roberts and Sherratt 2002; Jansen and van Baalen 2006; and Traulsen and Schuster 2003), and examples have begun to show up in species as diverse as amoebas (Queller et al. 2003), yeast (Smukalla et al. 2008), and fire ants (Keller and Ross 1998). Green beard theory may even help explain how mammalian mothers and fetuses interact (Haig 1996; Summers and Crespi 2005). The intellectual payoff of green beard theory for the study of human interactions outside the womb, however, is less clear. Price has attempted to solve the problem of a weak link between green beards and actual cooperativeness by suggesting that cooperativeness itself were to serve as a green beard. The result is a set of predictions very similar to those derived from reciprocity theory regarding cooperative partner choice and from indirect reciprocity theory regarding the importance of reputation.

30. Ostrom 1990.
31. Ehrhart and Keser 1999.
32. Page et al. 2005.
33. A good place to start reading the large literature on this topic is Cosmides and Tooby (2005).
34. Wason 1966.
35. You need to turn over the D and 7 cards. The D card is obvious, and almost everyone gets it. The 7 card is trickier. You need to turn it over to check whether there is a D on the other side. The rule says nothing about the K card, so you don't need to turn it over. Many people want to turn over the 3 card because it is mentioned in the rule, but it is not necessary. The rule states only that if there is a D on one side, then there must be a 3 on the other. It does not say that if there is a 3 on one side then it must have a D on the other.
36. Sugiyama et al. 2002.
37. Gigerenzer and Hug 1992.
38. Henrich and Henrich 2007:142–43.
39. Boyd and Richerson 1988.
40. Trivers 2006. See also Takezawa and Price (2010).
41. In order to distinguish between reciprocity and indirect reciprocity, many writers now refer to "direct reciprocity." We resist this move on the grounds that "reciprocity" is a perfectly good description of the phenomenon in question (see Box 4.2). Technically, "indirect reciprocity" is not reciprocity at all. Although we

are tempted to encourage people to adopt some alternative name for it, we fear that ship has already sailed.

42. Barry 1999.

43. Alexander 1977, 1987.

44. Nowak and Sigmund 2005.

45. Boyd and Richerson 1989.

46. Nowak and Sigmund 1998.

47. Wedekind and Milinski 2000.

48. See Bshary and Grutter (2006) for a nonhuman example of image scoring.

49. Sugden 1986; Leimar and Hammerstein 2001; Panchanathan and Boyd 2003.

50. Leimar and Hammerstein 2001; Panchanathan and Boyd 2003.

51. Panchanathan and Boyd 2004.

52. Milinski et al. 2002.

53. See, for example, Chong (1991). For a recent review of the literature on reputation management, see Tennie et al. (2010).

54. The Greater Internet Fuckwad Theory originated in an online comic called Penny Arcade (www.penny-arcade.com). We first heard about it while listening to Marketplace, a radio program produced by American Public Media, on August 2, 2010. The quote is from Eva Galperin of the Electronic Frontier Foundation, a civil liberties organization.

55. Bateson et al. 2006.

56. Ernest-Jones et al. 2011.

57. Burnham and Hare 2007.

58. Haley and Fessler 2005.

59. Rigdon et al. 2009.

60. Zhong et al. 2010.

61. Shariff and Norenzayan 2007.

62. Hardy and Van Vugt 2006.

63. Kurzban et al. 2007.

64. Smith 2003, 2010.

65. Dunbar 1997.

66. The only possible exceptions of which we are aware are recruitment signals in ants and bees, and possibly also ravens (see Bickerton 2009 for more on displacement, animal signaling systems, and the evolution of language).

67. See, for example, Bickerton (2009); see also Cronk (2004a).

68. Henrich and Henrich 2007.

69. Henrich and Henrich 2007:59.

70. Henrich and Henrich 2007:232.

71. Smith 2003, 2010.

72. Irons 2001; Cronk 2005.

73. Maynard Smith and Harper 2003.

74. Thapar 1989.

75. Guilford and Dawkins 1991; Grafen 1990a, 1990b.

76. Zahavi 1975.

77. Spence 1973.

78. Organ 1988; Deutsch Salamon and Deutsch 2006.
79. Kollman 1998.
80. Cronk 2005.
81. Boone 1998; Smith 2003.
82. Gintis et al. 2001.
83. Boone 1998; Harbaugh 1998; Hardy and Van Vugt 2006; Roberts 1998; Lotem et al. 2003; and Van Vugt et al. 2007.
84. Zahavi and Zahavi 1997:125–50.
85. Barclay and Willer 2007; see also Sylwester and Roberts 2010.
86. Boas 1966; Rosman and Rubel 1971.
87. Bliege Bird et al. 2001; Smith and Bliege Bird 2000.
88. Smith et al. 2003.
89. Hawkes 1990, 1991, 1993.
90. Smith and Bliege Bird 2000; see also Hawkes and Bliege Bird 2002.
91. Schelling 1960.
92. Nesse 2001.
93. This is a reference to 1 Corinthians 1:18-25. The original Greek words *skandalon* and *moros*, from which the English words "scandalous" and "moronic" are derived, appear in most English translations of the Bible as "stumbling block" and "foolishness" (Strong 1890).
94. Irons 2001.
95. Sosis and Ruffle 2003, 2004.
96. Soler 2008.
97. Frank 1988.

CHAPTER 5
COOPERATION AND ORGANIZATION

1. Hardin 1968:1244.
2. Lewontin 1972.
3. For example, studies of cooperative tendencies in twins show less variation among identical twins than among fraternal twins; see Alford et al. (2005) and Fowler et al. (2008).
4. See Crawford and Ostrom (1995) for a valiant attempt to systematize the myriad definitions and types of institutions.
5. Huntington 1968:12.
6. Riker 1980:432.
7. Fukuyama 2011:451.
8. Crawford and Ostrom 1995:582.
9. Searle 2005.
10. See also Fodor (1974:103).
11. North 1990:3.
12. Fukuyama 2011.
13. Putnam et al. 1993.
14. Putnam et al. 1993:94.
15. Putnam et al. 1993:115.

16. Venkatesh 2006.

17. Tarrow 1996. For a discussion of these opportunity structures and their role in political protest, see Meyer (2004).

18. McAdam et al. 1997:155.

19. Lichbach 1998; for a review of this concept, see Benford and Snow (2000).

20. Kubik 1994.

21. Rosenberg 1991.

22. For studies of the civil rights movement that informed this discussion, see McAdam (1982), Chong (1991), Chappell (1994), and McAdam et al. (1997).

23. Netting 1981.

24. Walston 1988:98.

25. Cronk et al. 2009.

26. Cronk and Steadman 2002.

27. And in the South Side neighborhood that Venkatesh studied—where the park was indeed a kind of common-pool resource, since gang use of the park took away from recreational use of the park—regular meetings were set up to monitor the agreement and deal with any new issues.

28. For more on the importance of the cultural context for common-pool resource management schemes, see McCay and Jentoft (1998).

29. See, for example, Wilson (1973).

30. North 1990.

31. Baumgartner and Jones 1993, 2002; Jones and Baumgartner 2005.

32. For an example of a nineteenth-century evolutionist, see Morgan (1877); on structural-functionalism, see Malinowski (1944) and Radcliffe-Brown (1957); on neofunctionalism, see Harris (1979) and Rappaport (1968).

33. Campbell 1975; Hayek 1978. For more on Hayek's cultural group selectionism, see Steele (1987).

34. Boyd and Richerson 1985; Henrich 2004; Henrich and Henrich 2007. See also Soltis et al. (1995).

35. Boyd and Richerson 1985; Bentley et al. 2004; cf. Eriksson and Coultas 2009.

36. Hallpike (1995) objects to this idea on the grounds that cultural group selection should be applied only to "self-sufficient communities." We see no reason to limit the theory's application in this way. Indeed, given how much most human societies have relied on trade for at least the past several millennia (Ridley 2011), we are unsure what a truly "self-sufficient community" would look like.

37. This shift was the unintended consequence of British colonial policies that led to high rates of intermarriage between the Yaaku-speaking Mukogodo and Maasai and other Maa-speaking pastoralists. For the full story, see Cronk (2004b).

38. Henrich and Henrich 2007; Cosmides and Tooby 2005.

39. Henrich 2004.

40. Chudek and Henrich 2011.

41. Sethi and Somanathan 2004.

42. From the *Seinfeld* episode titled "The Label Maker."

43. Bacharach 2006; Sugden 1993, 2009.

44. Patton 2000.

45. Alexander 1990; Flinn et al. 2005; Flinn and Alexander 2007; Flinn and Coe 2007.

46. Levine et al. 2005. For more on sports fandom and coalitional psychology, see Winegard and Deaner (2010).

47. Kurzban et al. 2001.

48. Puurtinen and Mappes 2009.

49. Bernhard et al. 2006.

50. Patton 2000, 2004, 2005.

51. Gil-White 2001.

52. Cronk 2002, 2004b.

53. Weber 1976; Robb 2007.

CHAPTER 6
MEETING AT PENN STATION

1. See, for example, Alvard (2001), Chwe (2001), Ahdieh (2010), Bicchieri (1993), Skyrms (2004), Binmore (1994), Sugden (2004), Bergstrom (2002), McAdams (2008).

2. Ullmann-Margalit 1977:129–30.

3. Cooper 1999.

4. Powell 2009, emphasis added.

5. Lewis 1969; Bicchieri 1993; Chwe 2001.

6. Kobayashi and Kohshima 1997, 2001.

7. See Wyman and Tomasello (2007).

8. Emery 2000.

9. For example, Trevarthen and Aitken (2005).

10. Baron-Cohen 1995; Baron-Cohen et al. 1995; Emery 2000.

11. Hrdy 2009.

12. Hagen and Bryant 2003.

13. Wiltermuth and Heath 2009.

14. Cohen et al. 2010.

15. Maynard Smith and Szathmáry 1997. Peter Corning (2005) also describes the Sculling Game, but he changes the payoffs so that it is no longer equivalent to the Prisoner's Dilemma.

16. Hume 1740: pt. 2, sec. 2.

17. Bicchieri 2006. See also Benvenisti (2002) and Bacharach (2006).

18. Elster 1979:146.

19. McAdams 2008:22–23.

20. Chong 1991:82.

21. Rousseau [1775] 1992:47.

22. Chong 1991; McAdam 1982; Chappell 1994.

23. For economic discussions of strategic voting, see Blair (1981) or Dasgupta and Maskin (2004). On a related point, see Tullock (1975).

24. Baumgartner et al. 2009. Larry Bartels (1988:58) has made a similar observation about the U.S. presidential primary process: potential candidates can-

not get attention in the media unless they have become front-runners, and they cannot become front-runners unless they get attention from the media.

25. Maynard Smith and Parker 1976.
26. Davies 1978.
27. Chwe 2001.
28. Fisher 1956.
29. Scott 1992.
30. Van Huyck et al. 1990, 1997.
31. Chaudhuri et al. 2009.
32. Barth 1969.
33. McElreath et al. 2003.
34. Ensminger 1997:7.
35. Seabright 2004.
36. Efferson et al. 2008.
37. Alvard 2003b; Alvard and Nolin 2002; Barnes 1996.
38. Sheehan 1985.
39. Van den Berghe 1979.
40. Nolin 2010.
41. Lansing 2006.
42. Lansing and Miller 2005. At first glance, this may seem counterintuitive: Isn't it the upstream farmer who suffers from having to share water? No. Because the upstream farmer is upstream, he never has to worry about water scarcity. Because of the threat of pests, the upstream farmer always has an incentive to plant at the same time as the downstream farmer. The set of incentives faced by the downstream farmer, on the other hand, depends upon how much pests reduce his crop yields compared to how much sharing water reduces them. If pests are bad enough, then both farmers will favor simultaneous cropping.
43. Sugden 2004.
44. Gudeman 1986.
45. Aktipis et al. 2011.
46. See, for example, Spencer (1965:27, 59) and Spencer (1988:39).
47. See, for example, Hollis ([1905] 1971:289, 321–22).
48. Jacobs 1965:210.
49. Cronk 2007; Cronk and Wasielewski 2008.
50. See also Spencer (1965:59).
51. See also Spencer (1965:27) and Perlove (1987).
52. Priming effects differ from framing effects in that the former are more at a gut level, shaping emotional responses rather than providing specific scripts that guide behavior. For more on priming effects, see Oyserman and Lee (2007).

CHAPTER 7
COOPERATION EMERGENT

1. Epstein and Axtell 1996:35.
2. For example, Alexander (1920); see Epstein (2006).
3. For example, Epstein (2006).
4. Lewes 1890:412.

5. Goldstein 1999:49.
6. Sawyer 2005:2.
7. Epstein and Axtell 1996:35.
8. Miller and Page 2007:46.
9. Christakis and Fowler 2009:26.
10. Cronk 1988.
11. Hume 1740: pt. 2, sec. 6.
12. Ferguson [1767] 1966:122.
13. Smith [1776] 1936:423.
14. Polanyi 1951.
15. For example, Hayek (1973).
16. Hayek 1945.
17. Sugden 1986, 1989.
18. Sumpter 2006; Couzin et al. 2005.
19. Seeley 2002.
20. Seeley 1998.
21. Krugman 1996.
22. Epstein 2006:5.
23. For example, van der Leeuw and Kohler (2007); anyone interested in the use of agent-based models in the study of cooperation should also read Miller and Page (2007).
24. Lansing 1993; Lansing and Miller 2005; Lansing 2006.
25. Menger 1892. Menger's verbal model has been formalized by Jones (1976).
26. Thurnwald 1912, 1934.
27. Blackwood 1935.
28. Oliver 1955.
29. Note that none of these examples rests in any way on conformist cultural transmission as theorized by Boyd and Richerson (1985). Although a tendency to mimic others might accelerate the rate at which a norm or convention emerges, these scenarios do not require it. They all assume that the actors involved make their choices solely on rational choice grounds.
30. Krugman 1996; Elster 1979, 1983.
31. Ullmann-Margalit 1997.
32. The essay appeared as a postscript in Hayek's *The Constitution of Liberty* (1960).
33. Ullmann-Margalit 1978:268.
34. Hayek 1960:159. See also Sugden (2004:57).
35. Barrett 2000; Guthrie 1993.
36. Stated in a debate with Natalie Angier at SUNY Albany in 2007; see http://www.edge.org. Wilson's misunderstanding of the invisible hand metaphor is also evident in his book *Darwin's Cathedral* (Wilson 2002:240).
37. Ullmann-Margalit 1997.
38. Nozick 1974; Elster 1979, 1983; Ullmann-Margalit 1977, 1978, 1997.
39. Ullmann-Margalit 1977.
40. Young 1993, 1998; Sugden 1989, 2004; Bicchieri 2006.
41. Ullmann-Margalit 1978:271.
42. For example, Elster (1979).

43. Langlois 1986.

44. But it does not make them right. Some phenomena may appear to be the spontaneous products of human action while actually having been deliberately designed. One possible example is "Astroturf lobbying," which refers to a lobbying campaign that has been designed to look like a spontaneous "grassroots" effort but that was actually designed by an interested party (Kollman 1998). Some refer to these as "hidden hand" explanations (Nozick 1974).

45. Flynn et al. 2009. See also Borroz (2009).

46. Schelling 1969.

47. Schelling's simple model has spawned a large amount of subsequent research. For a review, see Clark and Fossett (2008). Schelling's logic can also be applied to other situations, such as seating choices in a classroom (Ruoff and Schneider 2006).

48. Young 1998.

49. Sawyer 2003.

50. For more on the central limit theorem and emergence, see Miller and Page (2007:46–48).

51. Sumpter 2006:11.

52. Couzin 2009; see also Miller and Page 2007:46–48.

53. Gutenberg and Richter 1954.

54. Technically, we should say "power law-like curves." True power law curves go on forever, which is why gravity is a good example. Many phenomena simply cannot go on forever, but their curves still closely resemble true power law curves.

55. Zipf 1949.

56. Bentley and Maschner 2001; Bentley 2008; Hahn and Bentley 2003.

57. Herzog et al. 2004.

58. See, for example, Bikhchandani et al. (1992), Jones et al. (2003), Granovetter (1978), Granovetter and Soong (1988), Kim and Bearman (1997), Lohmann (1994), Oliver and Maxwell (2001).

59. Kuran 1991a:121.

60. Kuran 1989:41.

61. Kuran 1989:44; Schapiro 1984:19.

62. Scott 1992.

63. Tocqueville [1856] 1955:177; quoted in Kuran 1991b:24.

64. Biggs 2005.

65. In the case of earmarks, there is often less need for individual lobbyists to jump on a bandwagon because the good being pursued is private rather than public.

66. Baumgartner et al. 2009; Baumgartner and Leech 2001.

CHAPTER 8
MEETING AT GRAND CENTRAL

1. Hirt et al. 2008.
2. Ostrom 1990.

3. Hirt et al. 2008.

4. Hardin 1968.

5. Ostrom 1990; Glick 1970; Maass and Anderson 1986.

6. Paciotti et al. 2005; Paciotti and Borgerhoff Mulder 2004; Paciotti and Hadley 2003, 2004.

7. Curtin 1972.

8. Jones 1987.

9. This account is based on the Wikipedia page on the disaster (http://en.wikipedia.org/wiki/Tenerife_disaster; accessed on September 22, 2010).

10. Wilson 1998.

11. Cronk 1999.

12. By contrast, Wilson predicts that the social sciences will split, "with one part folding into or becoming continuous with biology, the other fusing with the humanities" (Wilson 1998:12) We see a continued role for systematic social science, independent from but consilient with biological sciences.

13. Bedau 1997, 2003.

14. See Fodor (1974).

15. Dennett 1995:50.

16. Cosmides et al. 1992:13–14.

17. Lowie [1917] 1929:66.

18. Aunger 1994; Durham 1991; Cullen 1999 (255–61).

19. Hayek 1973:20–21.

20. Hayek 1979:155, emphasis in original. Not long after Hayek wrote those words, French anthropologist Claude Lévi-Strauss wrote something very similar: "Culture is neither natural nor artificial. It stems from neither genetics nor rational thought, for it is made up of rules of conduct, which were not invented and whose function is generally not understood by the people who obey them" (1985:34). Despite the similarity, Lévi-Strauss cited Hayek in neither the French (Lévi-Strauss 1983) nor the English version of his book.

21. Durkheim [1895] 1964:104.

22. Durkheim [1895] 1964:102.

23. Interestingly, family relationships were the subject matter of what some consider the first important work in evolutionary psychology, John Bowlby's *Attachment and Loss* (1969), although the label was not popularized until years later. Readers interested in how evolutionary theory can be used to better understand relationships within the family should read not only Bowlby but also Sarah Blaffer Hrdy's *Mother Nature* (1999) and *Mothers and Others* (2009).

24. D'Andrade 1992; Cronk 1999, 2007; Cronk and Wasielewski 2008.

25. Boyd and Richerson 1985.

26. Chong 1991.

27. Moe 1980b.

28. Southwell 2004.

29. Richerson and Boyd 1999.

30. Haselton and Buss 2000; Haselton and Nettle 2006; Nesse 2001.

31. Complete citations are provided in the individual chapters.

References

Aberle, David F. 1987. Distinguished lecture: What kind of science is anthropology? *American Anthropologist* 89(3):551–66.

Ahdieh, Robert. 2010. The visible hand: Coordination functions of the regulatory state. *Minnesota Law Review* 95:578–649.

Ainsworth, Scott H. 2002. *Analyzing Interest Groups: Group Influence on People and Policies*. New York: Norton.

Aktipis, C. Athena. 2004. Know when to walk away: Contingent movement and the evolution of cooperation. *Journal of Theoretical Biology* 231(2):249–60.

———. 2011. Is cooperation viable in mobile organisms? Simple Walk Away rule favors the evolution of cooperation in groups. *Evolution and Human Behavior* 32(4):263–76.

Aktipis, C. Athena, Lee Cronk, and Rolando de Aguiar. 2011. Risk-pooling and herd survival: An agent-based model of a Maasai gift-giving system. *Human Ecology* 39:131–40.

Alexander, Richard D. 1977. Natural selection and the analysis of human sociality. In *Changing Scenes in the Natural Sciences* (1776–1976 Bicentennial Symposium Monograph), ed. C. E. Goulden, 283–37. Philadelphia: Philadelphia Academy of Natural Sciences.

———. 1987. *The Biology of Moral Systems*. Hawthorne, N.Y.: Aldine.

———. 1990. *How Did Humans Evolve? Reflections on the Uniquely Unique Species*. Special Publication 1:1–38. Ann Arbor: University of Michigan Museum of Zoology.

Alexander, S. 1920. *Space, Time, and Deity: The Gifford Lectures at Glasgow, 1916–1918*. New York: Dover.

Alford, John R., Carolyn L. Funk, and John R. Hibbing. 2005. Are political orientations genetically transmitted? *American Political Science Review* 99:153–67.

Alford, John R., and John R. Hibbing. 2008. The new empirical biopolitics. *Annual Review of Political Science* 11:183–203.

Alvard, Michael. 2001. Mutualistic hunting. In *Meat-Eating and Human Evolution*, ed. Craig Stanford and Henry Bunn, 261–78. New York: Oxford University Press.

———. 2003a. The adaptive nature of culture. *Evolutionary Anthropology* 12(3):136–49.

———. 2003b. Kinship, lineage identity, and an evolutionary perspective on the structure of cooperative big game hunting groups in Indonesia. *Human Nature* 14(2):129–63.

Alvard, M., and D. Nolin. 2002. Rousseau's whale hunt? Coordination among big game hunters. *Current Anthropology* 43(4):533–59.

Ambrose, Stephen. 1992. *Band of Brothers*. New York: Simon & Schuster.

Amenta, Edwin. 2006. *When Movements Matter*. Princeton, N.J.: Princeton University Press.

Atran, S., and A. Norenzayan. 2004. Religion's evolutionary landscape: Counterintuition, commitment, compassion, communion. *Behavioral and Brain Sciences* 27:713–70.

Aumann, Robert J. 1981. Survey of repeated games. In *Essays in Game Theory and Mathematical Economics in Honor of Oskar Morgenstern*, ed. Robert J. Aumann, 11–42. Mannheim, Germany: Bibliographisches Institut.

Aunger, Robert. 1994. Are food avoidances maladaptive in the Ituri Forest of Zaire? *Journal of Anthropological Research* 50:277–310.

Avilés, L. 1993. Interdemic selection and the sex ratio: A social spider perspective. *American Naturalist* 142:320–45.

Axelrod, Robert. 1984. *The Evolution of Cooperation*. New York: Basic Books.

———. 1997. *The Complexity of Cooperation*. Princeton, N.J.: Princeton University Press.

Axelrod, Robert, and William D. Hamilton. 1981. The evolution of cooperation. *Science* 211:1390–96.

Bacharach, Michael. 2006. *Beyond Individual Choice: Teams and Frames in Game Theory*. Princeton, N.J.: Princeton University Press.

Barash, David P. 2003. *The Survival Game*. New York: Henry Holt/Times Books.

Barclay, P., and R. Willer. 2007. Partner choice creates competitive altruism in humans. *Proceedings of the Royal Society: Biological Sciences* 274:749–53.

Bardsley, Nicholas. 2008. Dictator game giving: altruism or artifact? *Experimental Economics* 11:122–33.

Barnes, R. 1996. *Sea Hunters of Indonesia*. Oxford: Clarendon.

Barkow, Jerome H., Leda Cosmides, and John Tooby, eds. 1992. *The Adapted Mind: Evolutionary Psychology and the Generation of Culture*. Oxford: Oxford University Press.

Baron-Cohen, S. 1995. *Mindblindness: An Essay on Autism and Theory of Mind*. Cambridge, Mass.: MIT Press.

Baron-Cohen, S., R. Campbell, A. Karmiloff-Smith, J. Grant, and J. Walker. 1995. Are children with autism blind to the mentalistic significance of the eyes? *British Journal of Developmental Psychology* 13:379–98.

Barrett, Justin L. 2000. Exploring the natural foundations of religion. *Trends in Cognitive Science* 4:29–34.

Barrett, Lisa Feldman, Kristen A. Lindquist, and Maria Gendron. 2007. Language as context for the perception of emotion. *Trends in Cognitive Science* 11(8):327–32.

Barry, Dave. 1999. *Dave Barry Turns 50*. New York: Ballantine.

Bartels, Larry. 1988. *Presidential Primaries and the Dynamics of Public Choice*. Princeton, N.J.: Princeton University Press.

Barth, Fredrik, ed. 1969. *Ethnic Groups and Boundaries*. London: Allen and Unwin.

Bateson, Melissa, Daniel Nettle, and Gilbert Roberts. 2006. Cues of being watched enhance cooperation in a real-world setting. *Biology Letters* 2(3):412–14.

Baumard, Nicolas. 2011. Punishment is not a group adaptation: Humans punish to restore fairness rather than to support group cooperation. *Mind and Society* 10(1):1–26.

Baumgartner, Frank R., Jeffrey M. Berry, Marie Hojnacki, David C. Kimball, and

Beth L. Leech. 2009. *Lobbying and Policy Change: Who Wins, Who Loses, and Why*. Chicago: University of Chicago Press.

Baumgartner, Frank R., and Bryan D. Jones. 1993. *Agendas and Instability in American Politics*. Chicago: University of Chicago Press.

———, eds. 2002. *Policy Dynamics*. Chicago: University of Chicago Press.

Baumgartner, Frank R., and Beth L. Leech. 1998. *Basic Interests: The Importance of Groups in Politics and in Political Science*. Princeton, N.J.: Princeton University Press.

———. 2001. Issue niches and policy bandwagons: Patterns of interest group involvement in national politics. *Journal of Politics* 63:1191–1213.

Bedau, M. A. 1997. Weak emergence. *Philosophical Perspectives* 11:375–99.

———. 2003. Downward causation and autonomy in weak emergence. *Principia* 6:5–50.

Benford, Robert D., and David A. Snow. 2000. Framing processes and social movements: An overview and assessment. *Annual Review of Sociology* 26:611–39.

Bentley, R. A. 2008. Random drift versus selection in academic vocabulary. *PLoS ONE* 3(8):e3057.

Bentley, R. A., M. W. Hahn, and S. J. Shennan. 2004. Random drift and culture change. *Proceedings of the Royal Society B* 271:1443–50.

Bentley, R. A., and H.D.G. Maschner. 2001. Stylistic change as a self-organized critical phenomenon? *Journal of Archaeological Method and Theory* 8:35–65.

Benvenisti, Eyal. 2002. *Sharing Transboundary Resources: International Law and Optimal Resource Use*. Cambridge: Cambridge University Press.

Berg, Joyce, John Dickhaut, and Kevin McCabe. 1995. Trust, reciprocity, and social history. *Games and Economic Behavior* 10:122–42.

Bergstrom, Theodore C. 2002. Evolution of social behavior: Individual and group selection. *Journal of Economic Perspectives* 16(2):67–88.

Bernhard, Helen, Urs Fischbacher, and Ernst Fehr. 2006. Parochial altruism in humans. *Nature* 442:912–15.

Bicchieri, Cristina. 1993. *Rationality and Coordination*. Cambridge: Cambridge University Press.

———. 2006. *The Grammar of Society: The Nature and Dynamics of Social Norms*. Cambridge: Cambridge University Press.

Bickerton, Derek. 2009. *Adam's Tongue*. New York: Hill and Wang.

Biggs, Michael. 2005. Strikes as forest fires: Chicago and Paris in the late nineteenth century. *American Journal of Sociology* 110(6):1684–1714.

Bikhchandani, Sushil, David Hirschleifer, and Ivo Welch. 1992. A theory of fads, fashion, custom, and cultural change as informational cascades. *Journal of Political Economy* 100(5):992–1026.

Binmore, Ken. 1994. *Game Theory and Social Contract*. Cambridge, Mass.: MIT Press.

———. 2007. *Game Theory: A Very Short Introduction*. Oxford: Oxford University Press.

Blackwood, B. 1935. *Both Sides of Buka Passage*. Oxford: Clarendon.

Blair, Douglas H. 1981. On the ubiquity of strategic voting opportunities. *International Economic Review* 22:649–55.

Bliege Bird, R., E. A. Smith, and D. W. Bird. 2001. The hunting handicap: Costly signaling in male foraging strategies. *Behavioral Ecology and Sociobiology* 50:9–19.

Blurton Jones, N. G. 1984. Selfish origin for human food sharing: Tolerated theft. *Ethology and Sociobiology* 5:1–3.

———. 1987. Bushman birth spacing: Direct tests of some simple predictions. *Ethology and Sociobiology* 8:183–204.

Boas, F. 1966. *Kwakiutl Ethnography*. Chicago: University of Chicago Press.

Boehm, Christopher. 2008. Purposive social selection and the evolution of human altruism. *Cross-Cultural Research* 42(4):319–52.

Boone, James L. 1998. The evolution of magnanimity: When is it better to give than to receive? *Human Nature* 9(1):1–21.

Borroz, Tony. 2009. MIT hopes to exorcise "'phantom'" traffic jams. *Wired.com*.

Bourke, Andrew F. G. 2011. *Principles of Social Evolution*. Oxford: Oxford University Press.

Bowlby, J. 1969. *Attachment and Loss, vol. 1: Attachment*. New York: Basic Books.

Bowles, Samuel, and Herbert Gintis. 2003. Origins of human cooperation. In *Genetic and Cultural Evolution of Cooperation*, ed. P. Hammerstein, 429–43. Cambridge, Mass.: MIT Press.

Boyd, Robert, and Peter J. Richerson. 1985. *Culture and the Evolutionary Process*. Chicago: University of Chicago Press.

———. 1988. The evolution of reciprocity in sizable groups. *Journal of Theoretical Biology* 132:337–56.

———. 1989. The evolution of indirect reciprocity. *Social Networks* 11:213–36.

Brannon, E. M. 2005. Quantitative thinking: From monkey to human and human infant to adult. In *From Monkey Brain to Human Brain*, ed. S. Dehaene, J. Duhamel, M. D. Hauser, and G. Rizzolatti, 97–116. Cambridge, Mass.: MIT Press.

Brown, J. L. 1983. Cooperation—A biologist's dilemma. In *Advances in the Study of Behaviour*, ed. J. S. Rosenblatt, 1–37. New York: Academic Press.

Brown, W. M., and C. Moore. 2000. Is prospective altruist-detection an evolved solution to the adaptive problem of subtle cheating in cooperative ventures? Supportive evidence using the Wason selection task. *Evolution and Human Behavior* 21:25–37.

Brown, W. M., B. Palameta, and C. Moore. 2003. Are there nonverbal cues to commitment? An exploratory study using the zero-acquaintance video presentation paradigm. *Evolutionary Psychology* 1:42–69.

Brown, William M., Lee Cronk, Amy Jacobson, Keith Grochow, C. Karen Liu, Zoran Popovic, and Robert Trivers. 2005. Dance reveals symmetry especially in young men. *Nature* 438:1148–50.

Bshary, Redouan, and Alexandra S. Grutter. 2006. Image scoring and cooperation in a cleaner fish mutualism. *Nature* 441:975–78.

Buchanan, James M., Robert D. Tollison, and Gordon Tullock, eds. 1980. *Toward a Theory of the Rent-Seeking Society*. College Station: Texas A&M University Press.

Burnham, T. C., and B. Hare. 2007. Engineering human cooperation: Does involuntary neural activation increase public goods contributions? *Human Nature* 18:88–108.

Burnham, T. C., and D.D.P. Johnson. 2005. The biological and evolutionary logic of human cooperation. *Analyse und Kritik* 27:113–35.

Burt, Austin, and Robert Trivers. 2008. *Genes in Conflict.* Cambridge, Mass.: Belknap.

Buss, D. M., M. Abbott, A. Angleitner, A. Biaggio, A. Blanco-Villasenor, M. Bruchon Schweitzer, et al. 1990. International preferences in selecting mates: A study of 37 societies. *Journal of Cross-Cultural Psychology* 21:5–47.

Buss, David M., Martie G. Haselton, Todd K. Shackelford, April L. Bleske, and Jerome C. Wakefield. 1998. Adaptations, exaptations, and spandrels. *American Psychologist* 53(5):533–48.

Campbell, Andrea Louise. 2003. Participatory reactions to policy threats: Senior citizens and the defense of Social Security and Medicare. *Political Behavior* 25:29–49.

———. 2005. *How Policies Create Citizens.* Princeton, N.J.: Princeton University Press.

Campbell, Donald T. 1975. On the conflicts between biological and social evolution and between psychology and moral tradition. *American Psychologist* 30:1103–26.

Campbell, Lorne, Lee Cronk, Jeffry Simpson, Alison Milroy, Carol Wigington, and Bria Dunham. 2009. The association between men's ratings of women as desirable long-term mates and individual differences in women's sexual attitudes and behaviors. *Personality and Individual Differences* 46:509–13.

Carneiro, R. L. 1970. A theory of the origin of the state. *Science* 169:733–38.

Carre, Justin M., and Cheryl M. McCormick. 2008. In your face: Facial metrics predict aggressive behavior in the laboratory and in varsity and professional hockey players. *Proceedings of the Royal Society B* 275:2651–56.

Carr-Saunders, A. M. 1922. *The Population Problem: A Study in Human Evolution.* Oxford: Clarendon.

Cavalli-Sforza, L. L., and M. W. Feldman. 1981 *Cultural Transmission and Evolution.* Princeton, N.J.: Princeton University Press.

Chagnon, Napoleon A. 1975. Genealogy, solidarity and relatedness : Limits to local group size and patterns of fissioning in an expanding population. *Yearbook of Physical Anthropology* 19:5–110.

———. 1988. Male Yanomamo manipulations of kinship classifications of female kin for reproductive advantage. In *Human Reproductive Behavior: A Darwinian Perspective*, ed. L. Betzig, M. Borgerhoff Mulder, and P. Turke, 23–48. Cambridge: Cambridge University Press.

———. 2000. Manipulating kinship rules: A form of male Yanomamo reproductive competition. In *Adaptation and Human Behavior: An Anthropological Perspective*, ed. Lee Cronk, William Irons, and Napoleon A. Chagnon, 115–31. Hawthorne, N.Y.: Aldine.

Chappell, David L. 1994. *Inside Agitators.* Baltimore: Johns Hopkins University Press.

Chaudhuri, Ananish, Andrew Schotter, and Barry Sopher. 2009. Talking ourselves to efficiency: Coordination in inter-generational minimum effort games with private, almost common and common knowledge of advice. *The Economic Journal* 119:91–122.

Chepko-Sade, B. D., and T. J. Olivier. 1979. Coefficient of genetic relationship and the probability of intragenealogical fission in *Macaca mulatta. Behavioral Ecology and Sociobiology* 5:263–78.

Cherry, Todd L., Peter Frykblom, and Jason F. Shogren. 2002. Hardnose the dictator. *American Economic Review* 92(4):1218–21.

Chong, Dennis. 1991. *Collective Action and the Civil Rights Movement.* Chicago: University of Chicago Press.

———. 2000. *Rational Lives: Norms and Values in Politics and Society.* Chicago: University of Chicago Press.

Christakis, Nicholas A., and James H. Fowler. 2009. *Connected: The Surprising Power of Our Social Networks and How They Shape Our Lives.* New York: Little, Brown.

Chudek, Maciej, and Joseph Henrich. 2011. Culture-gene coevolution, norm-psychology and the emergence of human prosociality. *Trends in Cognitive Sciences* 15(5):218–26.

Chwe, Michael Suk-Young. 2001. *Rational Ritual: Culture, Coordination, and Common Knowledge.* Princeton, N.J.: Princeton University Press.

Clark, Peter B., and James Q. Wilson. 1961. Incentive systems: A theory of organizations. *Administrative Science Quarterly* 6:129–66.

Clark, William A. V., and Mark Fossett. 2008. Understanding the social context of the Schelling segregation model. *Proceedings of the National Academy of Science* 105:4109–14.

Clements, Kevin C., and David W. Stephens. 1995. Testing models of non-kin cooperation: Mutualism and the prisoner's dilemma. *Animal Behaviour* 50: 527–35.

Clutton-Brock, Tim. 2009. Cooperation between non-kin in animal societies. *Nature* 462:51–57.

Cohen, Emma E. A., Robin Ejsmond-Frey, Nicola Knight, and R.I.M. Dunbar. 2010. Rowers' high: Behavioural synchrony is correlated with elevated pain thresholds. *Biology Letters* 6(1):106–8.

Connor, R. C. 1986. Pseudoreciprocity: Investing in mutualism. *Animal Behaviour* 34:1562–66.

Cooper, Russell. 1999. *Coordination Games: Complementarities and Macroeconomics.* Cambridge: Cambridge University Press.

Corning, Peter A. 2005. *Holistic Darwinism: Synergy, Cybernetics, and the Bioeconomics of Evolution.* Chicago: University of Chicago Press.

Cosmides, Leda, and John Tooby. 2005. Neurocognitive adaptations designed for social exchange. In *The Handbook of Evolutionary Psychology,* ed. David M. Buss, 584–627. New York: John Wiley.

Cosmides, Leda, John Tooby, and Jerome H. Barkow. 1992. Introduction: Evolutionary psychology and conceptual integration. In *The Adapted Mind: Evolutionary Psychology and the Generation of Culture,* ed. Jerome H. Barkow, Leda Cosmides, and John Tooby, 3–15. Oxford: Oxford University Press.

Couzin, Iain D. 2009. Collective cognition in animal groups. *Trends in Cognitive Sciences* 13:36–43.

Couzin, Iain D., Jens Krause, Nigel R. Franks, and Simon A. Levin. 2005. Effective leadership and decision-making in animal groups on the move. *Nature* 433:513–16.

Crawford, Sue E. S., and Elinor Ostrom. 1995. A grammar of institutions. *American Political Science Review* 89(3):582–600.

Cronk, Lee. 1988. Spontaneous order analysis and anthropology. *Cultural Dynamics* 1(3):282–308.

———. 1989. Strings attached. *The Sciences* 29(3):2–4.

———. 1994. Group selection's new clothes. *Behavioral and Brain Sciences* 17(4):615–16.

———. 1995. Is there a role for culture in human behavioral ecology? *Evolution and Human Behavior* 16(3):181–205.

———. 1999. *That Complex Whole: Culture and the Evolution of Human Behavior*. Boulder, Colo.: Westview.

———. 2002. From true Dorobo to Mukogodo Maasai: Contested ethnicity in Kenya. *Ethnology* 41(1):27–49.

———. 2004a. Continuity, displaced reference, and deception. *Behavioral and Brain Sciences* 27(4):510–11.

———. 2004b. *From Mukogodo to Maasai: Ethnicity and Cultural Change in Kenya*. Boulder, Colo.: Westview.

———. 2005. The application of animal signaling theory to human phenomena: Some thoughts and clarifications. *Social Science Information/Information sur les Sciences Sociales* 44(4):603–20.

———. 2007. The influence of cultural framing on play in the trust game: A Maasai example. *Evolution and Human Behavior* 28:352–58.

Cronk, Lee, Drew Gerkey, and William Irons. 2009. Interviews as experiments: Using audience effects to examine social relationships. *Field Methods* 21:331–46.

Cronk, Lee, and Shannon Steadman. 2002. Tourists as a common-pool resource: A study of dive shops on Utila, Honduras. In *Economic Development: An Anthropological Approach*, Vol. 19 of the Society for Economic Anthropology Monographs, ed. Jeffrey Cohen and Norbert Dannhaeuser, 51–68. Walnut Creek, Calif.: AltaMira Press.

Cronk, Lee, and Helen Wasielewski. 2008. An unfamiliar social norm rapidly produces framing effects in an economic game. *Journal of Evolutionary Psychology* 6(4):283–308.

Cullen, B. 1999. Parasite ecology and the evolution of religion. In *The Evolution of Complexity*, ed. F. Heylighen, J. Bollen, and A. Riegler. Dordrecht: Kluwer.

Curtin, Philip D. 1972. *The Atlantic Slave Trade: A Census*. Madison: University of Wisconsin Press.

D'Andrade, Roy G. 1992. Schemas and motivation. In *Human Motives and Cultural Models*, ed. Roy G. D'Andrade and Claudia Strauss, 23–44. Cambridge: Cambridge University Press.

Darwin, Charles R. 1859. *On the Origin of Species*. London: John Murray.

Dasgupta, P., and Eric Maskin. 2004. The fairest vote of all. *Scientific American* 290:92–97.

Davies, N. B. 1978. Territorial defence in the speckled wood butterfly (Pararge aegeria), the resident always wins. *Animal Behaviour* 26:138–47.

Dawes, R. 1980. Social dilemmas. *Annual Review of Psychology* 31:169–93.

Dawkins, Richard. 1976. *The Selfish Gene.* New York: Oxford University Press.

———. 1983. *The Extended Phenotype: The Long Reach of the Gene.* Oxford: Oxford University Press.

De Dreu, Carsten K. W., Lindred L. Greer, Michel J. J. Handgraaf, Shaul Shalvi, Gerben A. Van Kleef, Matthijs Baas, Femke S. Ten Velden, Eric Van Dijk, and Sander W. W. Feith. 2010. The neuropeptide oxytocin regulates parochial altruism among humans. *Science* 328:1408–11.

De Dreu, Carsten K. W., Lindred L. Greer, Gerben A. Van Kleef, Shaul Shalvi, and Michael J. J. Handgraaf. 2011. Oxytocin promotes human ethnocentrism. *Proceedings of the National Academy of Sciences* 108(4):1262–66.

de Quervain, Dominique J.-F., Urs Fischbacher, Valerie Treyer, Melanie Schellhammer, Ulrich Schnyder, Alfred Buck, and Ernst Fehr. 2004. The neural basis of altruistic punishment. *Science* 305:1254–58.

DeBruine, Lisa M. 2002. Facial resemblance enhances trust. *Proceedings of the Royal Society B* 269:1307–12.

Decety, Jean, Philip L. Jackson, Jessica A. Sommerville, Thierry Chaminade, and Andrew N. Meltzoff. 2004. The neural bases of cooperation and competition: An fMRI investigation. *NeuroImage* 23:744–51.

Dennett, Daniel. 1995. *Darwin's Dangerous Idea.* New York: Simon & Schuster.

Deutsch Salamon, Sabrina, and Yuval Deutsch. 2006. OCB as a handicap: An evolutionary psychological perspective. *Journal of Organizational Behavior* 27:185–99.

Doebeli, Michael, and Christoph Hauert. 2005. Models of cooperation based on the prisoner's dilemma and the Snowdrift game. *Ecology Letters* 8:748–66.

Dugatkin, Lee Alan. 1997. *Cooperation among Animals: An Evolutionary Perspective.* Oxford: Oxford University Press.

Dugatkin, Lee Alan, and Jean-Guy J. Godin. 1993. Female mate copying in the guppy (*Poecilia reticulata*): Age-dependent effects. *Behavioral Ecology* 4(4):289–92.

Dugatkin, Lee Alan, and A. Sih. 1998. Evolutionary ecology and partner choice. In *Cognitive Ecology*, ed. R. Dukas, 379–403. Chicago: University of Chicago Press.

Dunbar, Robin. 1997. *Grooming, Gossip and the Evolution of Language.* Cambridge, Mass.: Harvard University Press.

Durham, W. H. 1991. *Coevolution.* Stanford, Calif.: Stanford University Press.

Durkheim, Emile. [1895] 1964. *The Rules of the Sociological Method,* trans. Sarah A. Solovay and John H. Mueller, ed. George E. G. Catlin. New York: Free Press.

Efferson, Charles, Rafael Lalive, and Ernst Fehr. 2008. The coevolution of cultural groups and ingroup favoritism. *Science* 321:1844–49.

Ehrhart, Karl-Martin, and Claudia Keser. 1999. Mobility and cooperation: On the run. *CIRANO Scientific Series* 99s-24:1–12.

Eisenegger, Christoph, Markus Naef, Romana Snozzi, Markus Heinrichs, and Ernst Fehr. 2010. Prejudice and truth about the effect of testosterone on human bargaining behaviour. *Nature* 463:356–61.

Elster, J. 1979. *Ulysses and the Sirens: Studies in Rationality and Irrationality.* Cambridge: Cambridge University Press.

———. 1983. *Explaining Technical Change: A Case Study in the Philosophy of Science.* Cambridge: Cambridge University Press.

Emerson, Alfred E. 1960. The evolution of adaptation in population systems. In *Evolution after Darwin*, vol. 1, ed. Sol Tax, 307–48. Chicago: University of Chicago Press.

Emery, N. J. 2000. The eyes have it: The neuroethology, function and evolution of social gaze. *Neuroscience and Biobehavioral Reviews* 24:581–604.

Ensminger, Jean. 1997. Transaction costs and Islam: Explaining conversion in Africa. *Journal of Institutional and Theoretical Economics* 153:4–29.

Epstein, Joshua M. 2006. *Generative Social Science: Studies in Agent-Based Computational Modeling.* Princeton, N.J.: Princeton University Press.

Epstein, Joshua M., and R. L. Axtell. 1996. *Growing Artificial Societies: Social Science from the Bottom Up.* Washington, D.C.: Brookings Institution Press.

Eriksson, K., and J. C. Coultas. 2009. Are people really conformist-biased? An empirical test and a new mathematical model. *Journal of Evolutionary Psychology* 7(1):5–21.

Ernest-Jones, Max, Daniel Nettle, and Melissa Bateson. 2011. Effects of eye images on everyday cooperative behavior: A field experiment. *Evolution and Human Behavior* 32(3):172–78.

Fager, Charles E. 1974. *Selma, 1965.* New York: Scribner.

Fehr, Ernst, and Urs Fischbacher. 2004. Third-party punishment and social norms. *Evolution and Human Behavior* 25:63–87.

Fehr, E., and J. Henrich. 2003. Is strong reciprocity a maladaptation. In *Genetic and Culture Evolution of Cooperation*, ed. Peter Hammerstein, 55–82. Cambridge, Mass.: MIT Press.

Ferguson, A. [1767] 1966. *An Essay on the History of Civil Society.* Edinburgh: Edinburgh University Press.

Fisher, Ronald A. 1956. *Statistical Methods and Scientific Inference.* New York: Hafner. Fodor

Flinn, Mark V., and Richard D. Alexander. 2007. Runaway social selection in human evolution. In *The Evolution of Mind: Fundamental Questions and Controversies*, ed. Steven W. Gangestad and Jeffry A. Simpson, 249–55. New York: Guilford.

Flinn, M. V., and K. Coe. 2007. The linked red queens of human cognition, reciprocity, and culture. In *The Evolution of Mind: Fundamental Questions and Controversies*, ed. S. W. Gangestad and J. A. Simpson, 339–47. New York: Guilford.

Flinn, M. V., D. C. Geary, and C. V. Ward. 2005. Ecological dominance, social competition, and coalitionary arms races: Why humans evolved extraordinary intelligence. *Evolution and Human Behavior* 26(1):10–46.

Flynn, M. R., A. R. Kasimov, J.-C. Nave, R. R. Rosales, and B. Seibold. 2009. Self-sustained nonlinear waves in traffic flow. *Physical Review E* 79:056113.

Fodor, Jerry A. 1974. Special sciences (or the disunity of science as a working hypothesis). *Synthese* 28(2):97–115.

Forsythe, Robert, Joel L. Horowitz, N. E. Savin, and Martin Sefton. 1994. Fairness in simple bargaining experiments. *Games and Economic Behavior* 6:347–69.

Foster, K. R., and H. Kokko. 2009. The evolution of superstitious and superstition-like behaviour. *Proceedings of the Royal Society of London* B 276:31–37.

Fowler, James H., Laura A. Baker, and Christopher T. Dawes. 2008. Genetic variation in political participation. *American Political Science Review* 102:233–48.

Fowler, James H., and Christopher T. Dawes. 2008. Two genes predict voter turnout. *Journal of Politics* 70:579–94.

Frank, M. C., D. L. Everett, E. Fedorenko, and E. Gibson. 2008. Number as a cognitive technology: Evidence from Pirahã language and cognition. *Cognition* 108:819–24.

Frank, Robert. 1988. *Passions within Reason: The Strategic Role of Emotions.* New York: Norton.

Frohlich, Norman, and Joe A. Oppenheimer. 1970. I get by with a little help from my friends. *World Politics* 23:104–20.

Fukuyama, Francis. 2011. *The Origins of Political Order: From Prehuman Times to the French Revolution.* New York: Farrar, Straus and Giroux.

Gardner, A. 2009. Adaptation as organism design. *Biology Letters* 5:861–64.

Gardner, A., and A. Grafen. 2009. Capturing the superorganism: A formal theory of group adaptation. *Journal of Evolutionary Biology* 22:659–71.

Gardner, A., and S. A. West. 2010. Greenbeards. *Evolution* 64:25–38.

Geertz, Clifford. 1973. *The Interpretation of Cultures.* New York: Basic Books.

Gelman, R., and C. R. Gallistel. 2004. Language and the origin of numerical concepts. *Science* 306:441–43.

Gerkey, Drew. 2010. *From State Collectives to Local Commons: Cooperation and Collective Action among Salmon Fishers and Reindeer Herders in Kamchatka, Russia.* Ph.D. dissertation, anthropology, Rutgers University.

Gigerenzer, G., and K. Hug. 1992. Domain-specific reasoning: Social contracts, cheating, and perspective change. *Cognition* 43:127–71.

Gil-White, F. J. 2001. Are ethnic groups biological "species" to the human brain? Essentialism in our cognition of some social categories. *Current Anthropology* 42(4):515–54.

Gintis, Herbert. 2000. Strong reciprocity and human sociality. *Journal of Theoretical Biology* 206:169–79.

———. 2009. *Game Theory Evolving.* 2nd ed. Princeton, N.J.: Princeton University Press.

Gintis, Herbert, Eric Alden Smith, and Samuel Bowles. 2001. Costly signaling and cooperation. *Journal of Theoretical Biology* 213:103–19.

Glick, T. F. 1970. *Irrigation and Society in Medieval Valencia.* Cambridge, Mass.: Harvard University Press.

Goldstein, Jeffrey. 1999. Emergence as a construct: History and issues. *Emergence* 1:49–72.

Gould, S. J., and E. S. Vrba. 1982. Exaptation: A missing term in the science of form. *Paleobiology* 8:4–15.

Grafen, A. 1990a. Biological signals as handicaps. *Journal of Theoretical Biology* 144:517–46.

———. 1990b. Sexual selection unhandicapped by the Fisher process. *Journal of Theoretical Biology* 144:473–516.

Granovetter, Mark. 1978. Threshold models of collective behavior. *American Journal of Sociology* 83(6):1420–43.

Granovetter, Mark, and Roland Soong. 1988. Threshold models of diversity: Chinese restaurants, residential segregation, and the spiral of silence. *Sociological Methodology* 18:69–104.

Gray, Virginia, and David Lowery. 1996. A niche theory of interest representation. *Journal of Politics* 59:91–111.

Guala, Francesco. 2011. Reciprocity: Weak or strong? What punishment experiments do (and do not) demonstrate. *Behavioral and Brain Sciences* 35:1–59.

Gudeman, Stephen. 1986. *Economics as Culture: Models and Metaphors of Livelihood*. London: Routledge and Kegan Paul.

Guilford, Tim, and Marian Stamp Dawkins. 1991. Receiver psychology and the evolution of animal signals. *Animal Behaviour* 42(1):1–14.

Gutenberg, B., and C. F. Richter. 1954. *Seismicity of the Earth and Associated Phenomena*. 2nd ed. Princeton, N.J.: Princeton University Press.

Güth, Werner, Rolf Schmittberger, and Bernd Schwarz. 1982. An experimental analysis of ultimatum bargaining. *Journal of Economic Behavior and Organization* 3:367–88.

Guthrie, Stewart E. 1993. *Faces in the Clouds: A New Theory of Religion*. Oxford: Oxford University Press.

———. 2001. Why gods? A cognitive theory. In *Religion in Mind: Cognitive Perspectives on Religious Belief, Ritual and Experience*, ed. J. Andresen, 94–111. Cambridge: Cambridge University Press.

Hagen, Edward H., and Gregory A. Bryant. 2003. Music and dance as a coalition signaling system. *Human Nature* 14:21–51.

Hahn, M. W., and R. A. Bentley. 2003. Drift as a mechanism for cultural change: An example from baby names. *Proceedings of the Royal Society B* 270: S1–S4.

Haig, David. 1996. Gestational drive and the green-bearded placenta. *Proceedings of the National Academy of Sciences* 93:6547–51.

Haley, Kevin J., and Daniel M. T. Fessler. 2005. Nobody's watching? Subtle cues affect generosity in an anonymous economic game. *Evolution and Human Behavior* 26:245–56.

Hallpike, C. A. 1995. Comment on Soltis et al. (1995). *Current Anthropology* 36(3):484.

Hamilton, W. D. 1964. The genetical evolution of social behaviour I and II. *Journal of Theoretical Biology* 7:1–16, 17–52.

———. 1975. Innate social aptitudes of man: An approach from evolutionary genetics. In *Biosocial Anthropology*, ed. Robin Fox, 133–53. London: Malaby Press.

———. 1996. *Narrow Roads of Gene Land, I: Evolution of Social Behaviour*. Oxford: Freeman.

Hamlin, J. Kiley, Karen Wynn, and Paul Bloom. 2007. Social evaluation by pre-verbal infants. *Nature* 450:557–60.

Harbaugh, William T. 1998. The prestige motive for making charitable transfers. *American Economic Review* 88(2):277–82.

Harcourt-Smith, W. E., and L. C. Aiello. 2004. Fossils, feet, and the evolution of human bipedal locomotion. *Journal of Anatomy* 204(5):403–16.

Hardin, Garrett. 1968. The tragedy of the commons. *Science* 162:1243–48.

Hardin, Russell. 1982. *Collective Action*. Baltimore: Johns Hopkins University Press.

Hardy, Charlie L., and Mark Van Vugt. 2006. Nice guys finish first: The competitive altruism hypothesis. *Personality and Social Psychology Bulletin* 32:1402–13.

Harris, Marvin. 1979. *Cultural Materialism*. New York: Random House.

Haselhuhn, Michael P., and Barbara A. Mellers. 2005. Multiple perspectives on decision making: Emotions and cooperation in economic games. *Cognitive Brain Research* 23:24–33.

Haselton, Martie G., and David M. Buss. 2000. Error management theory: A new perspective on biases in cross-sex mind reading. *Journal of Personality and Social Psychology* 78:81–91.

Haselton, Martie G., and Daniel Nettle. 2006. The paranoid optimist: An integrative evolutionary model of cognitive biases. *Personality and Social Psychology Review* 10(1):47–66.

Hawkes, K. 1990. Why do men hunt? Some benefits for risky strategies. In *Risk and Uncertainty in Tribal and Peasant Economies*, ed. E. Cashdan, 145–66. Boulder, Colo.: Westview.

———. 1991. Showing off: Tests of an hypothesis about men's foraging goals. *Ethology and Sociobiology* 12:29–54.

———. 1993. Why hunter-gatherers work: An ancient version of the problem of public goods. *Current Anthropology* 34(4):341–61.

Hawkes, K., and R. Bliege Bird. 2002. Showing off, handicap signaling, and the evolution of men's work. *Evolutionary Anthropology* 11:58–67.

Hayek, Friedrich A. 1945. The use of knowledge in society. *American Economic Review* 35(4):519–30.

———. 1955. *The Counter-Revolution of Science*. New York: Free Press.

———. 1960. *The Constitution of Liberty*. Chicago: University of Chicago Press.

———. 1973. *Law, Legislation, and Liberty. Volume 1: Rules and Order*. Chicago: University of Chicago Press.

———. 1978. *New Studies in Philosophy, Politics, Economics and the History of Ideas*. London: Routledge.

———. 1979. *Law, Legislation and Liberty. Volume 3: The Political Order of a Free People*. Chicago: University of Chicago Press.

Henrich, Joseph. 2004. Cultural group selection, coevolutionary processes, and large-scale cooperation. *Journal of Economic Behavior and Organization* 53:3–35.

Henrich, Joseph, and Robert Boyd. 2001. Why people punish defectors. *Journal of Theoretical Biology* 208:79–89.

Henrich, N., and J. Henrich. 2007. *Why Humans Cooperate: A Cultural and Evolutionary Explanation*. Oxford: Oxford University Press.

Hermer-Vazquez, L., E. Spelke, and A. Katsnelson. 1999. Sources of flexibility in human cognition: Dual task studies of space and language. *Cognitive Psychology* 39:3–36.

Herzog, H. A., R. A. Bentley, and M. W. Hahn. 2004. Random drift and large shifts in popularity of dog breeds. *Proceedings of the Royal Society B* 271:S353–56.

Hirt, Paul, Annie Gustafson, and Kelli L. Larson. 2008. The mirage in the Valley of the Sun. *Environmental History* 13:482–514.

Hoffman, Elizabeth, Kevin McCabe, Keith Shachat, and Vernon Smith. 1994. Preferences, property rights, and anonymity in bargaining games. *Games and Economic Behavior* 7(3):346–80.

Hollis, A. C. [1905] 1971. *The Masai: Their Language and Folklore*. Freeport, N.Y.: Books for Libraries Press.

Hrdy, Sarah Blaffer. 1999. *Mother Nature: A History of Mothers, Infants, and Natural Selection*. New York: Pantheon.

———. 2009. *Mothers and Others: The Evolutionary Origins of Mutual Understanding*. Cambridge, Mass.: Harvard University Press.

Hume, David. 1740. *A Treatise of Human Nature. Book III: Of Morals*. London: Thomas Longman.

Humphrey, Nicholas. 1997. Varieties of altruism—and the common ground between them. *Social Research* 64:199–209.

Huntington, Samuel. 1968. *Political Order in Changing Societies*. New Haven, Conn.: Yale University Press.

Hurd, J. P. 1983. Kin relatedness and church fissioning among the "Nebraska" Amish of Pennsylvania. *Social Biology* 30:59–66.

Irons, W. 1998. Adaptively relevant environments versus the environment of evolutionary adaptedness. *Evolutionary Anthropology* 6(6):194–204.

———. 2001. Religion as a hard-to-fake sign of commitment. In *Evolution and the Capacity for Commitment*, ed. R. M. Nesse, 292–309. New York: Russell Sage.

Jacobs, A. 1965. *The Traditional Political Organization of the Pastoral Masai*. Ph.D. dissertation, University of Oxford.

Jansen, Vincent A., and Minus van Baalen. 2006. Altruism through beard chromodynamics. *Nature* 440:663–66.

Jasper, James M. 1998. The emotions of protest: Affective and reactive emotions in and around social movements. *Sociological Forum* 13:397–424.

———. 2011. Emotions and social movements: Twenty years of theory and research. *Annual Review of Sociology* 37:285–303.

Johnson, Dominic D. P. 2009. The error of god: Error management theory, religion, and the evolution of cooperation. In *Games, Groups, and the Global Good*, ed. Simon A. Levin, 169–80. Berlin: Springer-Verlag.

Johnson, Dominic D. P., Michael E. Price, and Masanori Takezawa. 2008. Renaissance of the individual: Reciprocity, positive assortment, and the puzzle of human cooperation. In *Foundations of Evolutionary Psychology*, ed. Charles Crawford and Dennis Krebs, 331–25. New York: Lawrence Erlbaum.

Johnson, Gary R. 1987. In the name of the fatherland: An analysis of kin term usage in patriotic speech and literature. *International Political Science Review* 8(2):165–74.

Jones, B. C., L. M. DeBruine, A. C. Little, R. P. Burriss, and D. R. Feinberg. 2007. Social transmission of face preferences among humans. *Proceedings of the Royal Society of London B* 274:899–903.

Jones, Bryan D., and Frank R. Baumgartner. 2005. *The Politics of Attention: How Government Prioritizes Problems.* Chicago: University of Chicago Press.

Jones, Bryan D., Tracy Sulkin, and Heather A. Larsen. 2003. Policy punctuations in American political institutions. *American Political Science Review* 97(1):151–69.

Jones, Howard. 1987. *Mutiny on the Amistad.* Oxford: Oxford University Press.

Jones, Robert A. 1976. The origin and development of media of exchange. *Journal of Political Economy* 84(4):757–75.

Keesing, Roger M. 1974. Theories of culture. *Annual Review of Anthropology* 3:73–97.

———. 1975. *Kin Groups and Social Structure.* New York: Holt, Rinehart and Winston.

Keller, Laurent, and Kenneth G. Ross. 1998. Selfish genes: A green beard effect in the red fire ant. *Nature* 394:573–75.

Kendal, R. L., J. R. Kendal, W. J. E. Hoppitt, and K. N. Laland. 2009. Identifying social learning in animal populations: A new "option-bias" method. *PLoS ONE* 4(8):e6541.

Kim, Hyojoung, and Peter S. Bearman. 1997. The structure and dynamics of movement participation. *American Sociological Review* 62(1):70–93.

King, David C., and Jack L. Walker. 1992. The provision of benefits by interest groups in the United States. *Journal of Politics* 54(2):394–426.

Kitts, David B. 1981. Review of Simpson (1980). *Quarterly Review of Biology* 56(4):457–58.

Knoke, David. 1988. Incentives in collective action organizations. *American Sociological Review* 53:311–29.

———. 1990. *Organizing for Collective Action: The Political Economies of Associations.* Hawthorne, N.Y.: Aldine.

Kobayashi, Hiromi, and Shiro Kohshima. 1997. Unique morphology of the human eye. *Nature* 387:767–68.

———. 2001. Unique morphology of the human eye and its adaptive meaning: Comparative studies on external morphology of the primate eye. *Journal of Human Evolution* 40:419–35.

Kollman, Ken. 1998. *Outside Lobbying: Public Opinion and Interest Group Strategies.* Princeton, N.J.: Princeton University Press.

Kollock, Peter. 1998. Social dilemmas: The anatomy of cooperation. *Annual Review of Sociology* 24:183–214.

Kosfeld, Michael, Markus Heinrichs, Paul J. Zak, Urs Fischbacher, and Ernst Fehr. 2005. Oxytocin increases trust in humans. *Nature* 435:673–76.

Krugman, Paul. 1996. *The Self-Organizing Economy.* Cambridge, Mass.: Wiley-Blackwell.

Kubik, Jan. 1994. *The Power of Symbols against the Symbols of Power: The Rise of Solidarity and the Fall of State Socialism in Poland.* State College: Penn State Press.

Kümmerli, Rolf, Maxwell N. Burton-Chellew, Adin Ross-Gillespie, and Stuart A. West. 2010. Resistance to extreme strategies, rather than prosocial preferences, can explain human cooperation in public goods games. *Proceedings of the National Academy of Science* 107(22):10125–30.

Kuran, Timur. 1989. Sparks and prairie fires: A theory of unanticipated political revolution. *Public Choice* 61(1):41–74.

———. 1991a. The East European revolution of 1989: Is it surprising that we were surprised? *American Economic Review* 81(2):121–25.

———. 1991b. The element of surprise in the East European revolution of 1989. *World Politics* 44(1):7–48.

Kurzban, R., and C. A. Aktipis. 2007. On detecting the footprints of multilevel selection. In *Evolution of Mind: Fundamental Questions and Controversies,* ed. S. W. Gangestad and J. A. Simpson, 226–32. New York: Guilford.

Kurzban, Robert, Peter DeScioli, and Erin O'Brien. 2007. Audience effects on moralistic punishment. *Evolution and Human Behavior* 28:75–84.

Kurzban, Robert, John Tooby, and Leda Cosmides. 2001. Can race be erased? Coalitional computation and social categorization. *Proceedings of the National Academy of Science* 98(26):15387–92.

Lack, David. 1954. *The Natural Regulation of Animal Numbers.* Oxford: Oxford University Press.

Laland, Kevin N., John Odling-Smee, and Sean Myles. 2010. How culture shaped the human genome: Bringing genetics and the human sciences together. *Nature Reviews Genetics* 11:137–48.

Laland, Kevin N., Kim Sterelny, John Odling-Smee, William Hoppitt, and Tobias Uller. 2011. Cause and effect in biology revisited: Is Mayr's proximate-ultimate dichotomy still useful? *Science* 334:1512–16.

Laland, Kevin N., and Kerry Williams. 1998. Social transmission of maladaptive information in the guppy. *Behavioral Ecology* 9(5):493–99.

Langlois, R. N. 1986. Rationality, institutions, and explanation. In *Economics as a Process: Essays in the New Institutional Economics,* ed. R. N. Langlois, 225–55. Cambridge: Cambridge University Press.

Lansing, J. Stephen. 1993. Emergent properties of Balinese water temples. *American Anthropologist* 95(1):97–114.

———. 2006. *Perfect Order: Recognizing Complexity in Bali.* Princeton, N.J.: Princeton University Press.

Lansing, J. Stephen, and John H. Miller. 2005. Cooperation, games, and ecological feedback: Some insights from Bali. *Current Anthropology* 46:328–34.

Ledyard, John O. 1995. Public goods: A survey of experimental research. In *Handbook of Experimental Economics,* ed. John H. Kagel and Alvin E Roth, 111–94. Princeton, N.J.: Princeton University Press.

Leech, Beth L. 2006. Funding faction or buying silence? Grants, contracts, and interest group lobbying behavior. *Policy Studies Journal* 34:17–35.

Lehmann, Laurent, Kevin R. Foster, Elhanan Borenstein, and Marcus W. Feldman. 2008. Social and individual learning of helping in humans and other species. *Trends in Ecology and Evolution* 23(12):664–71.

Leimar, Olof, and Peter Hammerstein. 2001. Evolution of cooperation through indirect reciprocity. *Proceedings of the Royal Society of London B* 268:745–53.

Levine, Mark, Any Prosser, David Evans, and Stephen Reicher. 2005. Identity and emergency intervention: How social group membership and inclusiveness of group boundaries shape helping behavior. *Personality and Social Psychology Bulletin* 31:443–53.

Lévi-Strauss, Claude. 1983. *Le Regard Eloigné*. Paris: Plon.

———. 1985. *The View from Afar*, trans. Joachim Neugroschel and Phoebe Hoss. New York: Basic Books.

Levitt, Steven D., and John A. List. 2007. What do laboratory experiments measuring social preferences reveal about the real world? *Journal of Economic Perspectives* 21(2):153–74.

Levy, Jacques E., and Cesar Chavez. 1975. *Autobiography of La Causa*. New York: Norton.

Lewes, George Henry. 1890. *Problems of Life and Mind*. London: Kegan Paul, Trench, and Trübner.

Lewis, David K. 1969. *Convention: A Philosophical Study*. Cambridge, Mass.: Harvard University Press.

Lewontin, Richard C. 1972. The apportionment of human diversity. *Evolutionary Biology* 6:381–98.

Lewontin, R. C., and L. C. Dunn. 1960. The evolutionary dynamics of a polymorphism in the house mouse. *Genetics* 45:705–22.

Lichbach, Mark Irving. 1994. What makes rational peasants revolutionary? Dilemma, paradox, and irony in peasant collective action. *World Politics* 46:383–418.

———. 1995. *The Rebel's Dilemma*. Ann Arbor: University of Michigan Press.

———. 1998. Contending theories of contentious politics and the structure-action problem of social order. *Annual Review of Political Science* 1:401–24.

Linden, David. 2007. *The Accidental Mind: How Brain Evolution Has Given Us Love, Memory, Dreams, and God*. Cambridge, Mass.: Harvard University Press.

Livingstone, Frank B. 1958. Anthropological implications of sickle cell gene distribution in West Africa. *American Anthropologist* 60:533–62.

Lloyd, William Forster. 1833. *Two Lectures on the Checks to Population*. Oxford: S. Collingwood.

Locke, John L. 2005. Looking for, looking at: Social control, honest signals and intimate experience in human evolution and history. In *Animal Communication Networks*, ed. Peter McGregor, 416–41. Cambridge: Cambridge University Press.

Lohmann, Susanne. 1994. The dynamics of informational cascades: The Monday demonstrations in Leipzig, East Germany, 1989–91. *World Politics* 47(1):42–101.

Lotem, Arnon, Michael A. Fishman, and Lewi Stone. 2003. From reciprocity to unconditional altruism through signalling benefits. *Proceedings of the Royal Society of London B* 270:199–205.

Low, Bobbi. 2001. *Why Sex Matters: A Darwinian Look at Human Behavior*. Princeton, N.J.: Princeton University Press.

Lowie, Robert. [1917] 1929. *Culture and Ethnology.* New York: Peter Smith.

Maass, A., and R. L. Anderson. 1986. . . . *And the Desert Shall Rejoice: Conflict, Growth and Justice in Arid Environments.* Malabar, Fla.: R. E. Krieger.

Malinowski, Bronislaw. 1944. *A Scientific Theory of Culture.* Chapel Hill: University of North Carolina Press.

Mange, Arthur P. 1964. Growth and inbreeding of a human isolate. *Human Biology* 36(2):104–33.

Marler, P., and M. Tamura. 1964. Culturally transmitted patterns of vocal behavior in sparrows. *Science* 146:1483–86.

Marlowe, Frank W., J. Colette Berbesque, Abigail Barr, Clark Barrett, Alexander Bolyanatz, Juan Camilo Cardenas, Jean Ensminger, Michael Gurven, Edwins Gwako, Joseph Henrich, Natalie Henrich, Carolyn Lesorogol, Richard McElreath, and David Tracer. 2008. More "altruistic" punishment in larger societies. *Proceedings of the Royal Society of London B* 275:587–90.

Marwell, Gerald, and Ruth E. Ames. 1979. Experiments on the provision of public goods. I. Resources, interest, group size, and the free rider problem. *American Journal of Sociology* 84:1335–60.

———. 1980. Experiments on the provision of public goods. II. Provision points, stakes, experiences, and the free rider problem. *American Journal of Sociology* 85:926–37.

———. 1981. Economists free ride, does anyone else? Experiments on the provision of public goods IV. *Journal of Public Economics* 15:295–310.

Matos, Ricardo J., and Ingo Schlupp. 2005. Performing in front of an audience: Signallers and the social environment. In *Animal Communication Networks,* ed. Peter McGregor, 63–83. Cambridge: Cambridge University Press.

Mauss, M. [1922] 1990. *The Gift: Forms and Functions of Exchange in Archaic Societies.* London: Routledge.

Maynard Smith, J. 1964. Group selection and kin selection. *Nature* 201:1145–47.

———. 1976. Group selection. *Quarterly Review of Biology* 51(2):277–83.

———. 1998. The origin of altruism (a review of Sober and Wilson 1998). *Nature* 393:639–40.

Maynard Smith, John, and David Harper. 2003. *Animal Signals.* Oxford: Oxford University Press.

Maynard Smith, J., and G. A. Parker. 1976. The logic of asymmetric contests. *Animal Behaviour* 24:159–75.

Maynard Smith, John, and Eörs Szathmáry. 1997. *The Major Transitions in Evolution.* Oxford: Oxford University Press.

Mayr, Ernst. 1961. Cause and effect in biology. *Science* 134:1501–6.

McAdam, Doug. 1982. *Political Process and the Development of Black Insurgency, 1930–1970.* Chicago: University of Chicago Press.

McAdam, Doug, John D. McCarthy, and Mayer N. Zald, eds. 1996. *Comparative Perspectives on Social Movements: Political Opportunities, Mobilizing Structures, and Cultural Framings.* New York: Cambridge University Press.

McAdam, Doug, Sidney Tarrow, and Charles Tilly. 1997. Toward an integrated perspective on social movements and revolution. In *Comparative Politics: Rationality, Culture, and Structure,* ed. Mark Irving Lichbach and Alan S. Zuckerman, 142–73. Cambridge: Cambridge University Press.

McAdams, Richard H. 2008. Beyond the prisoners' dilemma: Coordination, game theory and the law. John M. Olin Law and Economics Working Paper No. 437 (second series), Public Law and Legal Theory Working Paper No. 241. University of Chicago.

McCabe, Kevin, Daniel Houser, Lee Ryan, Vernon Smith, and Theodore Trouard. 2001. A functional imaging study of cooperation in two-person reciprocal exchange. *Proceedings of the National Academy of Science* 98:11832–35.

McCay, B. J., and S. Jentoft. 1998. Market or community failure? Critical perspectives on common property research. *Human Organization* 57:21–29.

McElreath, Richard, and Robert Boyd. 2007. *Mathematical Models of Social Evolution: A Guide for the Perplexed*. Chicago: University of Chicago Press.

McElreath, Richard, Robert Boyd, and Peter J. Richerson. 2003. Shared norms and the evolution of ethnic markers. *Current Anthropology* 44:122–29.

McGrew, W. C. 1992. *Chimpanzee Material Culture*. Cambridge: Cambridge University Press.

Mealey, L., C. Daood, and M. Krage. 1996. Enhanced memory for faces associated with potential threat. *Ethology and Sociobiology* 17(2):119–28.

Menger, C. 1892. On the origin of money. *Economic Journal* 2:239–55.

Meyer, David S. 2004. Protest and political opportunities. *Annual Review of Sociology* 30:125–45.

Milinski, Manfred, Dirk Semmann, and Hans-Jürgen Krambeck. 2002. Reputation helps solve the "tragedy of the commons." *Nature* 415:424–26.

Miller, John H., and Scott E. Page. 2007. *Complex Adaptive Systems: An Introduction to Computational Models of Social Life*. Princeton, N.J.: Princeton University Press.

Moe, Terry M. 1980a. A calculus of group membership. *American Journal of Political Science* 24:593–632.

———. 1980b. *The Organization of Interests: Incentives and the Internal Dynamics of Political Interest Groups*. Chicago: University of Chicago Press.

———. 1981. Toward a broader view of interest groups. *Journal of Politics* 43: 531–43.

Moore, G. E. 1903. *Principia Ethica*. London: Cambridge University Press.

Morgan, Lewis Henry. 1877. *Ancient Society*. London: Macmillan.

Muller, Edward, and Karl-Dieter Opp. 1986. Rational choice and rebellious collective action. *American Political Science Review* 80:471–89.

———. 1987. Rebellious collective action revisited. *American Political Science Review* 81:561–64.

Nesse, R. M. 2001. The smoke detector principle: Natural selection and the regulation of defensive responses. *Annals of the New York Academy of Science* 935:75–85.

———. 2005. Natural selection and the regulation of defenses: A signal detection analysis of the smoke detector principle. *Evolution and Human Behavior* 26:88–105.

———. 2007. Runaway social selection for displays of partner value and altruism. *Biological Theory* 2(2):143–55.

Netting, Robert McC. 1981. *Balancing on an Alp: Ecological Change and Con-*

tinuity in a Swiss Mountain Community. Cambridge: Cambridge University Press.

Newton-Fisher, Nicholas E. 2006. Female coalitions against male aggression in wild chimpanzees of the Budongo forest. *International Journal of Primatology* 27(6):1589–99.

Nikiforakis, Nikos. 2007. Punishment and counter-punishment in public good games: Can we really govern ourselves? *Journal of Public Economics* 92: 91–112.

Nolin, D. A. 2010. Food-sharing networks in Lamalera, Indonesia: Reciprocity, kinship, and distance. *Human Nature* 21(3):243–68.

North, Douglass C. 1990. *Institutions, Institutional Change and Economic Performance*. Cambridge: Cambridge University Press.

Nowak, Martin A. 2006. *Evolutionary Dynamics*. Cambridge, Mass.: Harvard University Press.

Nowak, Martin A., and Karl Sigmund. 1998. Evolution of indirect reciprocity by image scoring. *Nature* 393:573–77.

———. 2005. Evolution of indirect reciprocity. *Nature* 437:1291–98.

Nozick, R. 1974. *Anarchy, State, and Utopia*. New York: Basic Books.

Oates, Kerris, and Margo Wilson. 2002. Nominal kinship cues facilitate altruism. *Proceedings of the Royal Society B* 269:105–9.

Oda, Ryo. 1997. Biased face recognition in the prisoner's dilemma game. *Evolution and Human Behavior* 18:309–15.

Oda, Ryo, Takuya Naganawa, Shinsaku Yamauchi, Noriko Yamagata, and Akiko Matsumoto-Oda. 2009a. Altruists are trusted based on non-verbal cues. *Biology Letters* 5(6):752–54.

Oda, Ryo, Noriko Yamagata, Yuki Yabiku, and Akiko Matsumoto-Oda. 2009b. Altruism can be assessed correctly based on impression. *Human Nature* 20: 331–41.

Odling-Smee, F. J., K. N. Laland, and M. F. Feldman. 2003. *Niche Construction: The Neglected Process in Evolution*. Princeton, N.J.: Princeton University Press.

Okasha, Samir. 2006. *Evolution and the Levels of Selection*. Oxford: Oxford University Press.

———. 2010. Altruism researchers must cooperate. *Nature* 467:653–55.

Oliver, D. 1955. *A Solomon Island Society*. Boston: Beacon.

Oliver, Pamela E. 1984. "If you don't do it, nobody else will": Active and token contributors to local collective action. *American Sociological Review* 49: 601–10.

———. 1993. Formal models of collective action. *Annual Review of Sociology* 19:271–300.

Oliver, Pamela E., and Gerald Maxwell. 2001. Whatever happened to critical mass theory? A retrospective and assessment. *Sociological Theory* 19(3):292–311.

Oliver, Pamela E., and Daniel J. Myers. 2002. Formal models in studying collective action and social movements. In *Methods of Social Movement Research*, ed. Bert Klandermans and Suzanne Staggenborg, 32–61. Minneapolis: University of Minnesota Press.

Olson, Mancur. 1965. *The Logic of Collective Action*. Cambridge, Mass.: Harvard University Press.

Organ, D. W. 1988. *Organizational Citizenship Behavior: The Good Soldier Syndrome*. Lexington, Mass.: Lexington Books.

Ostrom, Elinor. 1990. *Governing the Commons: The Evolution of Institutions for Collective Action*. New York: Cambridge University Press.

———. 1998. A behavioral approach to the rational choice theory of collective action: Presidential Address, American Political Science Association, 1997. *American Political Science Review* 92:1–22.

———. 2003. Toward a behavioral theory linking trust, reciprocity, and reputation. In *Trust and Reciprocity: Interdisciplinary Lessons from Experimental Research*, ed. Elinor Ostrom and James Walker. 19–79. New York: Russell Sage.

Ostrom, Elinor, Roy Gardner, and James Walker. 1994. *Rules, Games, and Common-Pool Resources*. Ann Arbor: University of Michigan Press.

Oyserman, D., and S. W. S. Lee. 2007. Priming "culture": Culture as situated cognition. In *Handbook of Cultural Psychology*, ed. S. Kitayama and D. Cohen, 255–79. New York: Guilford.

Paciotti, Brian, and Monique Borgerhoff Mulder. 2004. Sungusungu: The role of preexisting and evolving social institutions among Tanzanian vigilante organizations. *Human Organization* 63:112–24.

Paciotti, Brian, and Craig Hadley. 2003. The ultimatum game in southwestern Tanzania: Ethnic variation and institutional scope. *Current Anthropology* 44:427–32.

———. 2004. Large-scale cooperation among Sungusungu "vigilantes" of Tanzania: Conceptualizing micro-economic and institutional approaches. *Research in Economic Anthropology* 23:117–45.

Paciotti, Brian, Craig Hadley, Christopher Holmes, and Monique Borgerhoff Mulder. 2005. Grass roots justice in Tanzania. *American Scientist* 98:58–65.

Page, Talbot, Louis Putterman, and Bulent Unel. 2005. Voluntary association in public goods experiments: Reciprocity, mimicry and efficiency. *Economic Journal* 115:1032–53.

Panchanathan, Karthik, and Robert Boyd. 2003. A tale of two defectors: The importance of standing for evolution of indirect reciprocity. *Journal of Theoretical Biology* 224:115–26.

———. 2004. Indirect reciprocity can stabilize cooperation without the second-order free rider problem. *Nature* 432:499–502.

Parker, Glenn R. 1996. *Congress and the Rent-Seeking Society*. Ann Arbor: University of Michigan Press.

Parsons, Talcott, and Robert F. Bales. 1955. *Family, Socialization and Interaction Process*. Glencoe, Ill.: Free Press.

Patton, John Q. 2000. Reciprocal altruism and warfare: a case from the Ecuadorian Amazon. In *Adaptation and Human Behavior: An Anthropological Perspective*, ed. L. Cronk, N. Chagnon, and W. Irons, 417–36. Hawthorne, N.Y.: Aldine.

———. 2004. Coalitional effects on reciprocal fairness in the ultimatum game: A case from the Ecuadorian Amazon. In *Foundations of Human Sociality:*

Ethnography and Experiments in 15 Small-Scale Societies, ed. J. Henrich, R. Boyd, S. Bowles, H. Gintis, and C. Camerer, 96–124. Oxford: Oxford University Press.

———. 2005. Meat sharing for coalitional support. *Evolution and Human Behavior* 26:137–57.

Perlove, Diane Catherine. 1987. *Trading for Influence: The Social and Cultural Economics of Livestock Marketing among the Highland Samburu of Northern Kenya*. Ph.D. dissertation, anthropology, University of California, Los Angeles.

Perry, George H., Nathaniel J. Dominy, Katrina G. Claw, Arthur S. Lee, Heike Fiegler, Richard Redon, John Werner, Fernando A. Villanea, Joanna L. Mountain, Rajeev Misra, Nigel P. Carter, Charles Lee, and Anne C. Stone. 2007. Diet and the evolution of human amylase gene copy number variation. *Nature Genetics* 39:1256–60.

Perry, S., M. Baker, L. Fedigan, K. Gros-Louis, K. Jack, K. MacKinnon, J. Manson, M. Panger, K. Pyle, and L. Rose. 2003. Social conventions in wild white-faced capuchin monkeys: Evidence for traditions in a neotropical primate. *Current Anthropology* 44:241–68.

Pfeiffer, Thomas, Claudia Rutte, Timothy Killingback, Michael Taborsky, and Sebastian Bonhoeffer. 2005. Evolution of cooperation by generalized reciprocity. *Proceedings of the Royal Society B* 272:1115–20.

Polanyi, Michael. 1951. *The Logic of Liberty: Reflections and Rejoinders*. Chicago: University of Chicago Press.

Powell, Michael. 2009. The morning after, Democrats regret lost chances to win. *New York Times*, November 5, A29.

Price, G. R. 1970. Selection and covariance. *Nature* 227:520–21.

Price, Michael E. 2006. Monitoring, reputation, and "greenbeard" reciprocity in a Shuar work team. *Journal of Organizational Behavior* 27:201–19.

———. 2008. The resurrection of group selection as a theory of human cooperation. *Social Justice Research* 21:228–40.

Putnam, Robert D., Robert Leonardi, and Raffaella Y. Nanetti. 1993. *Making Democracy Work: Civic Traditions in Modern Italy*. Princeton, N.J.: Princeton University Press.

Puurtinen, Mikael, and Tapio Mappes. 2009. Between-group competition and human cooperation. *Proceedings of the Royal Society B* 276:355–60.

Qirko, Hector N. 2001. The maintenance and reinforcement of celibacy in institutionalized settings. In *Anthropological Approaches to Sexual Celibacy*, ed. E. J. Sobo and S. Bell, 65–86. Madison: University of Wisconsin Press.

———. 2002. The institutional maintenance of celibacy. *Current Anthropology* 43(2):321–28.

———. 2004a. Altruistic celibacy, kin-cue manipulation, and the development of religious institutions. *Zygon: Journal of Religion and Science* 39(3):681–706.

———. 2004b. "Fictive kin" and suicide terrorism. *Science* 304(5667):50–51.

———. 2009. Altruism in suicide terror organizations. *Zygon: Journal of Religion and Science* 44(2):289–322.

Queller, D. C., E. Ponte, S. Bozzaro, and J. E. Strassmann. 2003. Single-gene greenbeard effects in the social amoeba, *Dictyostelium discoideum*. *Science* 299:105–6.

Quinlan, R., and M. Quinlan. 2008. Human lactation, pair-bonds and alloparents: A cross-cultural analysis. *Human Nature* 19(2):87–102.

Radcliffe-Brown, Alfred Reginald. 1957. *A Natural Science of Society*. Glencoe, Ill.: Free Press.

Rankin, Daniel J. 2011. The social side of Homo economicus. *Trends in Ecology and Evolution* 26(1):1–3.

Rappaport, Roy. 1968. *Pigs for the Ancestors*. New Haven, Conn.: Yale University Press.

Rawls, John. 1971. *A Theory of Justice*. Cambridge, Mass.: Harvard University Press.

Reader, Simon M., Matthew J. Bruce, and Susanne Rebers. 2008. Social learning of novel route preferences in adult humans. *Biology Letters* 4:37–40.

Rendell, L., and H. Whitehead. 2001. Culture in whales and dolphins. *Behavioral and Brain Sciences* 24(2):309–24.

Renz, Loren, Elizabeth Cuccaro, Leslie A. Marino, and Mirek Drozdzowski. 2002. *Giving in the Aftermath of 9/11: An Update on the Foundation and Corporate Response*. New York: Foundation Center.

Richerson, Peter J., and Robert Boyd. 1999. The evolutionary dynamics of a crude super organism. *Human Nature* 10:253–89.

Richerson, Peter J., Robert Boyd, and Joseph Henrich. 2003. The cultural evolution of human cooperation. In *The Genetic and Cultural Evolution of Cooperation*, ed. P. Hammerstein, 357–88. Cambridge, Mass.: MIT Press.

Ridley, Mark. 2001. *The Cooperative Gene: How Mendel's Demon Explains the Evolution of Complex Beings*. New York: Simon & Schuster.

Ridley, Matt. 2011. *The Rational Optimist: How Prosperity Evolves*. New York: Harper.

Rigdon, Mary, Keiko Ishii, Motoki Watabe, and Shinobu Kitayama. 2009. Minimal social cues in the dictator game. *Journal of Economic Psychology* 30:358–67.

Riker, William. 1980. Implications from the disequilibrium of majority rule for the study of institutions. *American Political Science Review* 74:432–46.

Rilling, James K., Andrea L. Glenn, Meeta R. Jairam, Giuseppe Pagnoni, David R. Goldsmith, Hanie A. Elfenbein, and Scott O. Lilienfeld. 2007. Neural correlates of social cooperation and non-cooperation as a function of psychopathy. *Journal of Biological Psychiatry* 61:1260–71.

Rilling, James K., David R. Goldsmith, Andrea L. Glenn, Meeta R. Jairam, Hanie A. Elfenbein, Julien E. Dagenais, Christina D. Murdock, and Giuseppe Pagnoni. 2008. The neural correlates of the affective response to unreciprocated cooperation. *Neuropsychologia* 46:1256–66.

Riolo, Rick L., Michael D. Cohen, and Robert Axelrod. 2001. Evolution of cooperation without reciprocity. *Nature* 414:441–43.

———. 2002. Reply to Roberts and Sherratt 2002. *Nature* 418:500.

Robb, Graham. 2007. *The Discovery of France*. New York: W. W. Norton.

Roberts, Gilbert. 1998. Competitive altruism: from reciprocity to the handicap principle. *Proceedings of the Royal Society B* 265(1394):427–31.

Roberts, G., and T. N. Sherratt. 2002. Does similarity breed cooperation? *Nature* 418:499–500.

Roney, J. R., K. N. Hanson, K. M. Durante, and D. Maestripieri. 2006. Reading men's faces: Women's mate attractiveness judgments track men's testosterone and interest in infants. *Proceedings of the Royal Society of London B* 273:2169–75.

Rosenberg, Gerald N. 1991. *The Hollow Hope: Can Courts Bring about Social Change?* Chicago: University of Chicago Press.

Rosman, A., and P. G. Rubel. 1971. *Feasting with Mine Enemy: Rank and Exchange among Northwest Coast Societies*. New York: Columbia University Press.

Rothenberg, Lawrence S. 1988. Organizational maintenance and the retention decision in groups. *American Political Science Review* 82:1129–52.

———. 1992. *Linking Citizens to Government: Interest Group Politics at Common Cause*. New York: Cambridge University Press.

Rousseau, Jean-Jacques. [1775] 1992. *Discourse on the Origin of Inequality*, trans. Donald A. Cress. Indianapolis: Hackett.

Rudé, George. 1981. *The Crowd in History: A Study of Popular Disturbances in France and England, 1730–1848*. London: Lawrence and Wishart.

Ruoff, Gabriele, and Gerald Schneider. 2006. Segregation in the classroom: An empirical test of the Schelling model. *Rationality and Society* 18:95–117.

Sahlins, Marshall. 1965. On the sociology of primitive exchange. In *The Relevance of Models for Social Anthropology*, ed. M. Banton, 139–236. London: Tavistock. Reprinted in *Stone Age Economics* (1972, Transaction).

Salisbury, Robert H. 1969. An exchange theory of interest groups. *Midwest Journal of Political Science* 13:1–32.

Sally, David. 1995. Conversation and cooperation in social dilemmas: A meta-analysis of experiments from 1958 to 1992. *Rationality and Society* 7:58–92.

Salmon, Catherine. 1988. The evocative nature of kin terminology in political rhetoric. *Politics and the Life Sciences* 17(1):51–57.

Sawyer, R. Keith. 2003. *Improvised Dialogues: Emergence and Creativity in Conversation*. Westport, Conn.: Greenwood.

———. 2005. *Social Emergence: Societies as Complex Systems*. Cambridge: Cambridge University Press.

Schapiro, L. 1984. *The Russian Revolutions of 1917: The Origins of Modern Communism*. New York: Basic Books.

Schelling, Thomas. 1960. *The Strategy of Conflict*. Cambridge, Mass.: Harvard University Press.

———. 1969. Models of segregation. *American Economic Review* 59:488–93.

Scott, James C. 1992. *Domination and the Arts of Resistance: Hidden Transcripts*. New Haven, Conn.: Yale University Press.

Seabright, Paul. 2004. *The Company of Strangers: A Natural History of Economic Life*. Princeton, N.J.: Princeton University Press.

Searle, John R. 2005. What is an institution? *Journal of Institutional Economics* 1(1):1–22.

Seeley, T. D. 1998. Thoughts on information and integration in honey bee colonies. *Apidologie* 29:67–80.

———. 2002. When is self-organization used in biological systems? *Biological Bulletin* 202:314–18.

Sell, Aaron, Leda Cosmides, John Tooby, Daniel Sznycer, Christopher von Rueden, and Michael Gurven. 2009. Human adaptations for the visual assessment of strength and fighting ability from the body and face. *Proceedings of the Royal Society B* 276:575–84.

Sethi, Rajiv, and E. Somanathan. 2004. What can we learn from cultural group selection and co-evolutionary models? *Journal of Economic Behavior and Organization* 53:105–8.

Shariff, Azim F., and Ara Norenzayan. 2007. God is watching you: Priming god concepts increases prosocial behavior in an anonymous economic game. *Psychological Science* 18(9):803–9.

Sheehan, Glenn W. 1985. Whaling as an organizing focus in northwestern Alaskan Eskimo society. In *Prehistoric Hunter Gatherers: The Emergence of Cultural Complexity*, ed. T. Douglas Price and James A. Brown, 123–53. New York: Academic Press.

Shermer, Michael. 1998. *Why People Believe Weird Things*. New York: Freeman.

Shweder, R. A. 2005. When cultures collide: Which rights? Whose tradition of values? A critique of the global anti-FGM campaign. In *Global Justice and the Bulwarks of Localism*, ed. C. L. Eisgruber and András Sajó, 181–99. Leiden: M. Nijhoff.

Simons, Marlise. 2012. 4 Kenyans to stand trial at Hague Court in 2008 violence. *New York Times*, January 24, A9.

Simpson, George Gaylord. 1980. *Why and How: Some Problems and Methods in Historical Biology*. Oxford: Pergamon.

Skyrms, Brian. 2004. *The Stag Hunt and the Evolution of Social Structure*. Cambridge: Cambridge University Press.

Smith, Adam. [1776] 1936. *An Inquiry into the Nature and Causes of the Wealth of Nations*. New York: Modern Library.

Smith, Eric Alden. 2003. Human cooperation: Perspectives from behavioral ecology. In *Genetic and Cultural Evolution of Cooperation*, ed. P. Hammerstein, 401–27. Cambridge, Mass.: MIT Press.

———. 2010. Communication and collective action: Language and the evolution of human cooperation. *Evolution and Human Behavior* 31:231–45.

Smith, Eric Alden, and Rebecca Bliege Bird. 2000. Turtle hunting and tombstone openings: Generosity and costly signaling. *Evolution and Human Behavior* 21:245–61.

Smith, Eric Alden, Rebecca Bliege Bird, and Doug Bird. 2003. The benefits of costly signaling: Meriam turtle hunters. *Behavioral Ecology* 14(1):116–26.

Smukalla, Scott, Marina Caldara, Nathalie Pochet, Anne Beauvais, Stephanie Guadagnini, Chen Yan, Marcelo D. Vinces, An Jansen, Marie Christine Prevost, Jean-Paul Latgé, Gerald R. Fink, Kevin R. Foster, and Kevin J. Verstrepen. 2008. *FLO1* is a variable green beard gene that drives biofilm-like cooperation in budding yeast. *Cell* 135(4):726–37.

Sober, Elliott, and David Sloan Wilson. 1998. *Unto Others: The Evolution and Psychology of Unselfish Behavior*. Cambridge, Mass.: Harvard University Press.

Soler, Montserrat. 2008. *The Faith of Sacrifice: Commitment and Cooperation in Candomblé, an Afro-Brazilian Religion*. Ph.D. dissertation, anthropology, Rutgers University.

——. In press. Costly signaling, ritual and cooperation: Evidence from Candomblé, an Afro-Brazilian religion. *Evolution and Human Behavior*.

Soltis, Joseph, Robert Boyd, and Peter J. Richerson. 1995. Can group-functional behaviors evolve by cultural group selection? An empirical test. *Current Anthropology* 36(3):473–94.

Sosis, Richard, and Bradley Ruffle. 2003. Religious ritual and cooperation: Testing for a relationship on Israeli religious and secular kibbutzim. *Current Anthropology* 44:713–22.

——. 2004. Ideology, religion, and the evolution of cooperation: Field tests on Israeli kibbutzim. *Research in Economic Anthropology* 23:89–117.

Southwell, Priscilla L. 2004. Five years later: A re-assessment of Oregon's vote by mail electoral process. *PS: Political Science & Politics* 37:89–93.

Spence, Michael. 1973. Job market signaling. *Quarterly Journal of Economics* 87(3):355–74.

Spencer, Paul. 1965. *The Samburu: A Study of Gerontocracy in a Nomadic Tribe*. Berkeley: University of California Press.

——. 1988. *The Maasai of Matapato: A Study of Rituals of Rebellion*. Bloomington: Indiana University Press in association with the International African Institute.

Steele, David Ramsay. 1987. Hayek's theory of cultural group selection. *Journal of Libertarian Studies* 8(2):171–95.

Strong, James. 1890. *A Concise Dictionary of the Words in the Greek Testament*. Madison, N.J. http://www.heraldmag.org/olb/contents/dictionaries/SGreek.pdf.

Sugden, Robert. 1986. *The Economics of Rights, Co-operation, and Welfare*. Oxford: Blackwell.

——. 1989. Spontaneous order. *Journal of Economic Perspectives* 3(4):85–97.

——. 1993. Thinking as a team: Toward an explanation of nonselfish behavior. *Social Philosophy and Policy* 10:69–89.

——. 2004. *The Economics of Rights, Co-operation, and Welfare*. 2nd ed. New York: Palgrave Macmillan.

——. 2009. Neither self-interest nor self-sacrifice: The fraternal morality of market relationships. In *Games, Groups, and the Global Good*, ed. Simon A. Levin, 259–83. Berlin: Springer-Verlag.

Sugiyama, L. S., J. Tooby, and L. Cosmides. 2002. Cross-cultural evidence of cognitive adaptations for social exchange among the Shiwiar of Ecuadorian Amazonia. *Proceedings of the National Academy of Sciences* 99(17):11537–45.

Summers, K., and B. J. Crespi. 2005. Cadherins in maternal-foetal interactions: Red queen with a green beard? *Proceedings Royal Society of London B* 272:643–49.

Sumpter, D. J. T. 2006. The principles of collective animal behavior. *Philosophical Transactions of the Royal Society of London B: Biological Sciences* 361:5–22.

Sutton, Robert I. 2010. *The No Asshole Rule: Building a Civilized Workplace and Surviving One That Isn't*. New York: Business Plus.

Swaddle, John P., and Innes C. Cuthill. 1994. Preference for symmetric males by female zebra finches. *Nature* 367:165–66.

Sylwester, Karolina, and Gilbert Roberts. 2010. Cooperators benefit through reputation-based partner choice in economic games. *Biology Letters* 6(5): 659–62.

Tabibnia, Golnaz, and Matthew D. Lieberman. 2007. Fairness and cooperation are rewarding: Evidence from social cognitive neuroscience. *Annals of the New York Academy of Sciences* 1118:90–101.

Takezawa, Masanori, and Michael Price. 2010. Revisiting "The evolution of reciprocity in sizable groups": Continuous reciprocity in the repeated N-person prisoner's dilemma. *Journal of Theoretical Biology* 264:188–96.

Tarrow, Sidney. 1996. States and opportunities: The political restructuring of movements. In *Comparative Perspectives on Social Movements*, ed. Doug McAdam, John D. McCarthy, and Mayer N. Zald, 41–61. New York: Cambridge University Press.

Tennie, Claudio, Uta Frith, and Chris D. Frith. 2010. Reputation management in the age of the world-wide web. *Trends in Cognitive Science* 14(11):482–88.

Terkel, Joseph. 1996. Cultural transmission of feeding behavior in the Black Rat (*Rattus rattus*). In *Social Learning in Animals: The Roots of Culture*, ed. Cecelia M. Heyes and Bennett G. Galef, 17–48. San Diego: Academic Press.

Thapar, Valmik. 1989. *Tiger: Portrait of a Predator*. New York: Facts on File.

Thurnwald, R. 1912. *Forschungen auf den Salomo-Inseln und dem Bismarck-Archipel*. Vol. 3. Berlin: Dietrich Reimer.

———. 1934. Pigs and currency in Buin: Observations about primitive standards of value and economics. *Oceania* 5(2):119–41.

Tillock, Harriet, and Denton E. Morrison. 1979. Group size and contribution to collective action: A test of Mancur Olson's theory on Zero Population Growth, Inc. *Research in Social Movements, Conflict, and Change* 2:131–58.

Tinbergen, Niko. 1963. On aims and methods in ethology. *Zeitschrift für Tierpsychologie* 20:410–33.

Tishkoff, S. A., Floyd A. Reed, Alessia Ranciaro, Benjamin F. Voight, Courtney C. Babbitt, Jesse S. Silverman, Kweli Powell, Holly Mortensen, Jibril B. Hirbo, Maha Osman, Muntaser Ibrahim, Sabah A. Omar, Godfrey Lema, T. B. Nyambo, Jilur Ghori, Suzannah Bumpstead, Jonathan K. Pritchard, Gregory A. Wray, and Panos Deloukas. 2007. Convergent adaptation of human lactase persistence in Africans and Europeans. *Nature Genetics* 39(1):31–40.

Tocqueville, Alexis de. [1856] 1955. *The Old Regime and the French Revolution*, trans. Stuart Gilbert. Garden City, N.Y.: Doubleday.

Tracer, D. 2003. Selfishness and fairness in economic and evolutionary perspective: An experimental economic study in Papua New Guinea. *Current Anthropology* 44:432–38.

Traulsen, Arne, and Heinz Georg Schuster. 2003. Minimal model for tag-based cooperation. *Physical Review E* 68(4):046129.

Trevarthen, Colwyn, and Kenneth J. Aitken. 2005. Infant intersubjectivity: Research, theory, and clinical applications. *Journal of Child Psychology and Psychiatry* 42:3–48.

Trivers, Robert. 1971. The evolution of reciprocal altruism. *Quarterly Review of Biology* 46(1):35–57.

———. 1985. *Social Evolution*. Menlo Park, Calif.: Benjamin Cummings.

———. 2004. Mutual benefits at all levels of life. *Science* 304:964–65.

———. 2006. Reciprocal altruism: 30 years later. In *Cooperation in Primates and Humans: Mechanisms and Evolution*, ed. P. M. Kappeller and C. P. van Schaik, 67–83. Berlin: Springer-Verlag.

Truman, David B. 1951. *The Governmental Process: Political Interests and Public Opinion*. New York: Knopf.

Tullock, Gordon. 1967. The welfare costs of tariffs, monopolies, and theft. *Western Economic Journal* 5:224–32.

———. 1975. The paradox of not voting for oneself. *American Political Science Review* 69:1295–97.

Tylor, E. B. 1871. *Primitive Culture*. New York: Henry Holt.

Ullmann-Margalit, Edna. 1977. *The Emergence of Norms*. Oxford: Clarendon.

———. 1978. Invisible-hand explanations. *Synthese* 39:263–91.

———. 1997. The invisible hand and the cunning of reason. *Social Research* 64:181–98.

United Nations, Office of the UN High Commissioner for Human Rights. 2008. Report from OHCHR Fact-Finding Mission to Kenya, February 6–28, 2008.

Van den Berghe, Pierre. 1979. *Human Family Systems: An Evolutionary View*. New York: Elsevier.

van der Leeuw, S., and T. Kohler, eds. 2007. *The Model-Based Archaeology of Socio-Natural Systems*. Santa Fe, N.M.: School of Advanced Research.

Van Huyck, John B., Raymond C. Battalio, and Richard Oliver Beil. 1990. Tacit coordination games, strategic uncertainty, and coordination failure. *American Economic Review* 80:234–48.

Van Huyck, John B., Raymond C. Battalio, and Frederick W. Rankin. 1997. On the origin of convention: Evidence from coordination games. *Economic Journal* 107:576–96.

Van Vugt, M., D. De Cremer, and D. Janssen. 2007. Gender differences in competition and cooperation: The male warrior hypothesis. *Psychological Science* 18:19–23.

Venkatesh, Sudhir Alladi. 2006. *Off the Books: The Underground Economy of the Urban Poor*. Cambridge, Mass.: Harvard University Press.

Walker, Jack L., Jr. 1983. The origins and maintenance of interest groups in America. *American Political Science Review* 77:390–406.

———. 1991. *Mobilizing Interest Groups in America*. Ann Arbor: University of Michigan Press.

Walston, James. 1988. *The Mafia and Clientelism: Roads to Rome in Post-War Calabria*. New York: Routledge.

Wason, P. C. 1966. Reasoning. In *New Horizons in Psychology*, ed. B. M. Foss, 135–51. Harmondsworth: Penguin.

Waynforth, D. 2007. Mate choice copying in humans. *Human Nature* 18(3):264–71.

Weber, Eugen. 1976. *Peasants into Frenchmen*. Stanford, Calif.: Stanford University Press.

Wedekind, Claus, and Manfred Milinski. 2000. Cooperation through image scoring in humans. *Science* 288:850–52.

West, Stuart A., Claire El Mouden, and Andy Gardner. 2011. Sixteen common misconceptions about the evolution of cooperation in humans. *Evolution and Human Behavior* 32(4):231–62.

West, S. A., A. S. Griffin, and A. Gardner. 2007. Social semantics: Altruism, cooperation, mutualism, strong reciprocity and group selection. *Journal of Evolutionary Biology* 20:415–32.

West-Eberhard, Mary Jane. 1975. The evolution of social behavior by kin selection. *Quarterly Review of Biology* 50(1):1–33.

———. 1979. Sexual selection, social competition, and evolution. *Proceedings of the American Philosophical Society* 51(4):222–34.

Williams, George C. 1966. *Adaptation and Natural Selection*. Princeton, N.J.: Princeton University Press.

Wilson, David Sloan. 2002. *Darwin's Cathedral*. Chicago: University of Chicago Press.

Wilson, David Sloan, and Elliott Sober. 1994. Reintroducing group selection to the human behavioral sciences. *Behavioral and Brain Sciences* 17(4):585–654.

Wilson, Edward O. 1998. *Consilience: The Unity of Knowledge*. New York: Knopf.

Wilson, Peter J. 1973. *Crab Antics: The Social Anthropology of English-Speaking Negro Societies of the Caribbean*. New Haven, Conn.: Yale University Press.

Wiltermuth, Scott S., and Chip Heath. 2009. Synchrony and cooperation. *Psychological Science* 20:1–5.

Winawer, J., N. Witthoft, M. Frank, L. Wu, A. Wade, and L. Boroditsky. 2007. The Russian Blues reveal effects of language on color discrimination. *Proceedings of the National Academy of Science* 104(19):7780–85.

Winegard, Benjamin, and Robert O. Deaner. 2010. The evolutionary significance of Red Sox Nation: Sport fandom as a by-product of coalitional psychology. *Evolutionary Psychology* 8(3):432–46.

Wyman, Emily, and Michael Tomasello. 2007. The ontogenetic origins of human cooperation. In *Oxford Handbook of Evolutionary Psychology*, ed. Robin Dunbar and Louise Barrett, 227–36. Oxford: Oxford University Press.

Wynne-Edwards, V. C. 1962. *Animal Dispersion in Relation to Social Behavior*. London: Oliver & Boyd.

Yamagishi, Toshi, Shigehito Tanidab, Rie Mashimab, Eri Shimomab, and Satoshi Kanazawa. 2003. You can judge a book by its cover: Evidence that cheaters may look different from cooperators. *Evolution and Human Behavior* 24:290–301.

Yorzinski, J. L., and M. L. Platt. 2010. Same-sex gaze attraction influences mate-choice copying in humans. *PLoS ONE* 5(2):e9115.

Yoshida, Wako, Ben Seymour, Karl J. Friston, and Raymond J. Dolan. 2010. Neural mechanisms of belief inference during cooperative games. *Journal of Neuroscience* 30:10744–51.

Young, H. Peyton. 1993. The evolution of conventions. *Econometrica* 61(1):57–84.

———. 1998. Conventional contracts. *Review of Economic Studies* 65(4):773–92.

Zahavi, Amotz. 1975. Mate selection—A selection for a handicap. *Journal of Theoretical Biology* 53:205–14.

Zahavi, Amotz, and Avishag Zahavi. 1997. *The Handicap Principle*. Oxford: Oxford University Press.

Zhong, Chen-Bo, Vanessa K. Bohns, and Francesca Gino. 2010. Good lamps are the best police: Darkness increases dishonesty and self-interested behavior. *Psychological Science* 21(3):311–14.

Zipf, G. K. 1949. *Human Behavior and the Principle of Least Effort*. Cambridge, Mass.: Addison-Wesley.

Index

AARP, 58, 194n14

adaptation, 18–46, 183–84; as compromise, 24; cooperation in relation to, 18–19; culture and, 31–34; defined, 11–12, 18; environmental role in, 27–30; fortuitousness vs., 12, 21–22, 38; judgment and, 30–31; language and, 34; levels of, 34–44; and levels of explanation, 19–20; moral considerations concerning, 44–45; natural selection and, 22–23; new uses for, 25–26; and phylogeny, 24–27; process of, 22; sociocultural processes of, 26

adaptively relevant environment (ARE), 27–28

agency, 30, 158

agenda setting, 133–34

agent-based models, 155–56

Ainsworth, Scott, 7

air traffic, 2, 174–75

Aktipis, Athena, 44, 81

Alexander, Richard, 75, 87, 119

altruism: in behavioral experiments, 42–43; competitive, 95–96; cooperation confused with, 7–8, 190n24; defined, 7; factors encouraging, 5; gene-level selection and, 39; identification of persons exhibiting, 83; kinship and, 26; natural selection and, 7–8; reciprocity and, 7–8, 76, 77

Alvard, Michael, 142

Amenta, Edwin, 63–65

American Political Science Association (APSA), 57–58

Americans Disabled for Accessible Public Transit (ADAPT), 68

Americans with Disabilities Act (1990), 68

Ames, Ruth, 66–67

Amin, Idi, 171

Amish, 41

La Amistad (ship), 2, 173–74

amniotes, 27–28

Anabaptists, 40–41

analysis of variance (ANOVA), 38

Anderson, Raymond, 170

animal behaviorism, 32

animals. *See* nonhumans

anonymity, behavior under conditions of, 89–90

anthropology, 31, 178–81

anti-coordination games, 134–35

Apple computer corporation, 137–38

apples, 21

Arabian babbler, 95–96

Arizona, 169–70

armadillos, 29

artificer bias, 158

artificial selection, 22–23

Association of Utila Dive Shop Owners (AUDSO), 112–14

assurance games, 132, 165

Astroturf lobbying, 204n44

attention, 127, 167–68

Au, 16

audience: effect of, on cooperative behaviors, 89–91; evolutionary, 43–44

Aumann, Robert, 78

avalanches (information), 165

Axelrod, Robert, 78, 81, 130; *The Evolution of Cooperation*, 78

Axtell, Robert, 153

babblers (birds), 95–96

Bacharach, Michael, 119

balanced reciprocity, 77

Bali, 143–44

bandwagons, 165, 167

Barash, David, 7

Barry, Dave, 87

Bartels, Larry, 201n24

barter, 156–57

Barth, Fredrik, 139

Battle of the Sexes Game, 132–33, 150, 167

Baumgartner, Frank, 167

Bay Islands Conservation Association, 112

bees, 155

behavioral experiments, 29–30, 42–44, 138–41. *See also* experimental economic games; *particular experiments by name*

behavioral sciences. *See* social and behavioral sciences
behavior genetics, 72–73
Bentley, R. Alexander, 164
Bernhard, Helen, 121–22
Bicchieri, Christina, 159
big game hunting, 96–97
Biggs, Michael, 166–67
Binmore, Ken, 6, 77–78
biological group selection, 34–44, 184
biology: areas of study in, 177; levels of explanation in, 12, 19–20, 176. *See also* evolutionary biology
biparental care, 24, 25, 82. *See also* cooperative breeding
Black Cat (gang leader), 105–6
Black Kings, 105–6
Bloomberg, Michael, 125
Bougainville, Melanesia, 157
Bourgeois strategy, 135
Bourke, Andrew F. G., 190n24
Bowlby, John, 205n23
Bowles, Samuel, 7
Boyd, Robert, 27, 86, 88, 92, 116, 139, 182, 203n29
brain, 25, 73
broadcast efficiency, 95
Brown v. Board of Education (1954), 107–8
Buin, 157
Burnham, Terrence, 43
Buss, David, 30
by-product mutualism, 21–22
by-product theory of public goods provisioning, 58–59

Calabria, Italy, 111
Campbell, Donald T., 116
Candomblé, 98–99
Carr-Saunders, Alexander, 35, 192n66
cascades (information), 165
categories, 49, 117
causal explanations, 6, 12, 20, 176, 190n2
central limit theorem, 163
Chagnon, Napoleon, 41
Charlie task, 127–28
Chaudhuri, Ananish, 138–39
Chavez, Cesar, 60
cheaters, detection of, 84–86, 185
chemistry, 177
Chicago: collective action in housing project in, 105–6; labor strikes in, 166–67

Chicken (game), 134–35, 150
Chong, Dennis, 61, 130–31, 132
Christakis, Nicholas, 153
Chwe, Michael, 13, 137, 165
Cinqué, Joseph, 173–74
cities, size of, 164
Civil Rights Act (1964), 107, 108
civil rights movement, 4, 61, 107–8, 131, 132, 181
Clark, Peter B., 59–61, 72, 73, 119
Clements, Robert Earl, 63
club goods. *See* toll goods
coalitional psychology, 118–23, 141, 185
Coe, Kathryn, 119
coercion, 51–52
collective action: defined, 9; entrepreneurs behind, 62–65; explanations for, 48; group size as factor in, 52, 56–57, 194n31; irrational behavior and, 69–70; patrons behind, 64–66
collective action dilemmas: conflicts of interest as factor in, 10, 183; coordination problems in relation to, 129–31; defined, 9–10; evolutionary approaches to, 5; game studying, 17; Olson on, 48, 50–52, 56–60; Prisoner's Dilemma as instance of, 79–80; rational self-interest approaches to, ix, 4, 48, 50–51; second-order, 51; strategies for overcoming, 4, 12, 51–52, 56–69, 130–31
collective animal behavior, 154–55, 163
collective goods: as by-product of membership groups, 57–59; characteristics of, 9; defined, 54. *See also* public goods
commitments, 97–98, 109
Common Cause, 66
common goods, 9. *See also* public goods
common knowledge: conventions as vehicle of, 136; coordination dependent on, 13, 125–28, 130, 132–33, 135–39, 166–68, 183, 185–86; cultural differences as obstacle to, 130; frames, scripts, and schemata as vehicles of, 144–45; hard-to-fake signals and, 94; metaknowledge of, 13, 128, 137, 165, 186; norms as vehicle of, 152; rituals as vehicle of, 137
common-pool resources, 9–10; excludability of, 53–54, 108; management of, 13, 108–15; public goods compared to, 108; subtractability of, 53, 101, 108; water as instance of, 169–71
communes, 99

community fissioning, 40–41
competitive altruism, 95–96
Conambo, Ecuador, 122
conflicts of interest, 10, 94, 132, 175
consciousness raising, 108
conservatism, 158
consilience, 6, 14, 175–83
conventions (behavior), 135–38, 142, 151–52, 163
cooperation: adaptation in relation to, 18–19; altruism confused with, 7–8, 190n24; biological factors in, 72–74; biparental care and, 24, 25; collective action dilemmas and, 9–10; cooperativeness distinguished from, 8; coordination in relation to, 10, 149; defined, 7–8; diverse examples of, 2–3; error management theory and, 30–31; kinship and, 26–27; language and, 34; organizations and, 13; paradox of, ix–x, 2; psychological factors in, 12–13; reciprocity in relation to, 86; study of, x, 3–7, 71, 175–76. See also collective action dilemma
cooperative breeding, 128. See also biparental care
cooperativeness, 8, 24
cooperators, identification of, 74, 77–78, 80–84, 185
co-opted adaptation, 25–26
coordination: anti-coordination games, 134–35; Balinese, 143–44; cooperation in relation to, 10, 149; defined, 10; emergence of, 14, 151–68; games studying, 150; government regulation of, 152; in laboratory experiments, 138–41; natural selection for, 128–29; pleasures of, 128–29; real-world, 141–44; and trust, 132
coordination problems, 13–14; collective action dilemmas in relation to, 129–31; defined, 10; examples of, 124–26; information's role in, 10, 13, 124–26, 130, 132–49, 166–68, 183, 185–86; solutions for, 125–26; solving, 10–11, 135–38, 141, 186; study of, 175–76
copying, 164. See also mimicking
Corning, Peter, 201n15
Cosmides, Leda, 84–85
costliness of signals, 93
crab antics, 115
critical mass, 165

Cronk, Lee, x, xi, 11, 111–15, 118, 136, 145–49
cues, 155
cultural anthropology, 180
cultural drift, 164
cultural frames, 106–8
cultural group selection, 13, 115–23
culture: adaptation and, 31–34; age of, 32–33; behavior in relation to, 180–81; defining, 31; explanatory value of, 31–32; gene-culture coevolution, 32–33, 85–86, 118–19; human, 33–34; nonhuman, 32

dairy products, 32–33
Darwin, Charles, 22–23, 158, 159, 187
Davies, Nick, 135
Dawkins, Richard, 36, 189n14, 196n29
DeBruine, Lisa, 27
delayed reciprocity, 75
Dennett, Daniel, 177
descent groups, 142–43. See also kinship
design: error as feature of, 30; flaws in, 24; invisible-hand explanations vs., 158–59, 204n44; in natural and artificial selection, 22–23
Deutsch, Yuval, 94
developmental explanations, 19
Dictator Games, 15, 16, 41, 42, 90
direct reciprocity, 75, 197n41
displacement, 91–92
distal explanations, 12, 19–20
dominance hierarchies, 36
driving, side-of-road norms for, 151–52, 160–61
Duchenne smiles, 83
Durkheim, Emile, 179–80

earthquakes, 164
economics, 3; experimental games in, 15–17, 41–43, 88, 138–41; rational self-interest approaches in, 48, 50
Ecuadorian Amazon, 122
education, 55
Efferson, Charles, 140–41
efficacy costs, 93
efficacy problem, 56–57, 69–70, 194n31
elderly, pensions for, 63–64
Elster, Jon, 158, 159
emergence, 186–87; and complementarity of explanations, 176–77, 179–80; concept and definitions of, 153; information dispersal and, 154–55, 163–68;

emergence (*cont.*)
  mathematical characteristics of, 163–68,
  187; of norms and conventions, 14,
  151–68; pros and cons of, 160–61; in
  social sciences, 154–56; types of phenomena exhibiting, 161–62; weak, 177
Emerson, Alfred E., 35, 192n66
Emilia-Romagna, Italy, 104–5
empathy, 18
enforcement, 109
Ensminger, Jean, 140
entrepreneurs. *See* interest group
  entrepreneurs
environment of evolutionary adaptedness
  (EEA), 27
environments: adaptation and, 27–30; artificial, 29–30; novel, 29, 43, 182; organisms' shaping of, 28
ephemeral emergents, 162
Epstein, Joshua, 153, 155
error management theory, 30–31, 182
ethnic groups, 117–18, 121–23, 139–41
evolutionary audience, 43–44
evolutionary biology: on altruism, 7–8;
  areas of study in, 19; causal explanations in, 6, 12, 20, 72–73; on cooperation, 7–8; and proximate mechanisms,
  72–73; on reciprocity, 7–8; social and
  behavioral sciences in relation to, 11,
  14, 72, 175–83, 187; theoretical foundations of, 3
evolutionary psychology, 90–91, 205n23
exaptations, 26
excludability, 9, 53–54
experimental economic games, 15–17, 41–
  43, 88, 138–41. *See also* behavioral experiments; *particular games by name*
explanation: confusion of levels of, 20;
  evolutionary, 44–45; levels of, 12, 19–
  20, 72, 176; moral considerations concerning, 44–45. *See also* causal
  explanations
expressive benefits, 60, 131, 194n20. *See
  also* purposive benefits
extinction, 37
eyes: behavioral effects of feeling watched
  by, 90–91; and coordination problems,
  126–27; as kludges, 25

faces: psychological/behavioral characteristics attributed on basis of, 83; represen-
  tations of, 90; trust experiments using,
  27
fairness, 15–16, 67
feet, 24, 25, 27
Fehr, Ernst, 140
Ferguson, Adam, 154, 179, 186
firefighting, 54–55
firms, 117
Fisher, Ronald A., 136
fissioning, community, 40–41
Flinn, Mark, 119
flying fish, 18
focal points, 10, 13, 136
folk psychology, 90–91
food, 29
food riots, 60–61
fortuitousness, 12, 21–22, 38
Fowler, James, 153
framing, 145, 147–49, 202n52
Frank, Robert, 43–44, 99
free riding: avoidance of individuals prone
  to, 84–85; behavior contrary to, ix, 66–
  67; coercion and, 51; as collective action
  problem, 9, 50, 54, 57; excludability
  and, 53; group size and, 52, 56; second-
  order problems with, 88–89
Fukuyama, Francis, 104
functional magnetic resonance imaging
  (fMRI), 73

Game of Chicken, 134–35, 150
gametes, 35
game theory, 6–7, 15–17. *See also* Prisoner's Dilemma
gene-culture coevolution, 32–33, 85–86,
  118–19
generalized reciprocity, 76
generative social science, 155
generosity as performance, 86–99
Generous Tit-for-Tat strategy, 81
gene-level selection, 35, 36, 39
Gintis, Herbert, 6, 7, 76
Glick, Thomas, 170
Gnau, 16
Goldstein, Jeffrey, 153
goods, types of, 53–56
Gould, Stephen Jay, 25–26
government. *See* states
Gramercy Park, New York City, 110
Grand Central Terminal, New York City,
  10, 13, 126

gravity, 164
Greater Internet Fuckwad Theory, 89, 198n54
green beard theory, 196n29
groups, types of, 49. *See also* biological group selection; collective action; collective action dilemmas; organizations
Guala, Francesco, 76
Gudeman, Steve, 145
Gutenberg-Richter relationship, 164

Hallpike, C. A., 200n36
Hamilton, William D., 26, 39, 44, 196n29
Hammerstein, Peter, 88
Hardin, Garrett, 170; "The Tragedy of the Commons," 101
Hardin, Russell, 7, 65
hard-to-fake signals, 92–99
Haselton, Martie, 30, 182
Hawk-Dove Game, 134–35, 150
Hawkes, Kristen, 97
Hayek, Friedrich, 3, 116, 154, 155, 158, 179, 205n20
health care, 55
Henrich, Joseph, 7, 85, 92, 116, 118
Henrich, Natalie, 7, 85, 92
hidden-hand explanations, 204n44
hidden transcript, 137, 166
historical sciences, 3
hormones, 73–74
Hrdy, Sarah Blaffer, 128
Hughes, Howard, 65–66
humans: community fissioning in, 40–41; eyes of, 126–27; institutions as characteristic of, 101; language as characteristic of, 34, 91–92; nonhuman characteristics shared by, 32–34; and novel environments, 29; power law curves in behavior of, 164; psychosocial skills of, 23, 28, 86–87, 127–28
Hume, David, 77–78, 129, 154, 186
hunting, 96–97
Hurd, James, 41
Hutterites, 40–41
hyperactive agency detection device, 158

identity, 140
image scoring, 87–88
imitation. *See* copying; mimicking
incentives, for collective action, 57–60
inclusive fitness theory, 26

indexical signals, 93
indirect reciprocity, 75, 86–92, 119, 197n41
individualism, 114–15
individuals, 72–100; cheater detection by, 84–86; cooperator identification by, 74, 77–78, 80–84; indirect reciprocity of, 86–93; signaling by, 93–99
information: coordination problems stemming from, 10, 13, 124–26, 130, 132–49, 166–68, 183, 185–86; culture as socially-transmitted, 31; dispersal of, as central to emergence, 154, 155, 163–68; mathematical properties of, 163–68; in revolutionary societies, 166
institutions: between-group variations in, 101–2; cooperation aided by, 11; definitions of, 102–3; as human characteristic, 101; organizations distinguished from, 103; "stickiness" of, 115
intentionality, 127
interest group entrepreneurs, 62–65, 195n31
interest groups, 49
International Criminal Court, 65
intersubjectivity, 127–28
Investment Games, 16
invisible hand, 154, 156–60, 186–87, 204n44
Iranian Revolution (1979), 166
irrational thoughts/behavior: and collective action, 69–70; as signaling commitment, 98–99
Islam, 140
isotuatin (umbilical cords; risk-pooling partners), 145–47
Italy, 104–5, 111
iteration: coordination problems and, 130; reciprocity and, 78, 80–81; trust and, 132

Jacobs, Alan, 145–46
Jevons, William, 156
job market signaling, 93–94
Johnson, Dominic, 30, 43
joint attention, 127
jointness of consumption, 54
jointness of production, 54
jointness of supply, 53
judgment, adaptation and, 30–31
justice, grassroots, 1, 171–72

Kenya, 65, 118, 145–46
keyboards, 161
Khomeini, Ruhollah, 166
King, David, 67
King, Martin Luther, Jr., 107
kin selection theory, 26
kinship, 26–27, 103–4. *See also* descent
    groups
Kitts, David B., 3
kludges, 24–25
Knoke, David, 60, 66
knowledge. *See* common knowledge; infor-
    mation; metaknowledge
Kollman, Ken, 94
Krugman, Paul, 155, 158
Kümmerli, Rolf, 42
Kuran, Timur, 166
Kurzban, Robert, 44, 120, 141

laboratory behavioral experiments. *See* be-
    havioral experiments
lactose absorption, 32–33
Laland, Kevin N., 190n2
Lalive, Rafael, 140
Lamalera, Lembata, Indonesia, 141–43
language: adaptation and, 34; cooperation
    and, 34; displacement as characteristic
    of, 91–92; methodology for studying,
    147; and reputation, 34, 91–92; word
    frequencies in, 164
Lansing, Steve, 143–44, 156
latent groups, 49, 57
Leech, Beth L., ix, xi, 11, 49, 134, 167
Lehmann, Laurent, 7
Leimar, Olof, 88
Lenin, Vladimir, 166
Levine, Mark, 119–20
Lévi-Strauss, Claude, 205n20
Lewes, George Henry, 153
liberalism, 158
Lichbach, Mark, 60–61, 64, 67
Linden, David, 25
Lloyd, William Forster, 9, 101
lobbying, 58–59, 167–68, 204n44
long-term pair bonds, 25, 82–83
love, 99

Maasai, 118, 145–49
Maa speakers, 118
Maass, Arthur, 170
malaria, 33
Manchester United, 119–20

Mange, Arthur P., 40
Mappes, Tapio, 121
markets: common-pool resource manage-
    ment by, 109; as cooperative system, 19;
    cultural group selection in, 116–17
Marwell, Gerald, 66–67
Marx, Karl, 154
mate choice, 93
mate preferences, 25
material benefits, 60–62
mathematics, of emergent behaviors, 163–
    68, 187
Maynard Smith, John, 37, 39–40, 44, 129,
    134
Mayr, Ernest, 19
McAdams, Richard H., 130
McElreath, Richard, 139
meiosis, 35, 40
membership groups, 57–58, 60–63
Menger, Carl, 156–57, 159, 162, 187
mentalizing, 127–28
Mer turtle hunting, 96
metaknowledge, 13, 128, 136, 137, 165,
    186
metaphors, 145–46
methodologies, 6
Microsoft Word, 161
migration, 37
Milinski, Manfred, 88, 89
Miller, John H., 144, 153
mimicking, 32, 33, 203n29. *See also*
    copying
Minnesota-Dakotas Hardware Association,
    69
Minnesota Farm Bureau, 69
Minnesota Farmers Union, 69, 70
Minnesota Retail Federation, 69
mobilizational structures, 106–8
Moe, Terry, 61, 69–70, 181–82
money, 156–57, 159, 162, 177–78
monitoring, 109, 113
morality, evolutionary explanations and,
    44–45, 184
Morrison, Denton, 47
Mukogodo, 118, 122
Muller, Edward, 66
multilevel selection, 38, 44
mutual monitoring. *See* monitoring

National Organization on Disabilities
    (NOD), 68
National Rifle Association, 59

naturalistic fallacy, 45, 184
natural selection: adaptation and, 22–23; and altruism, 7–8; artificial vs., 22–23; biological group-level, 34–44; as invisible-hand explanation, 158; levels of, 5, 34–44; multilevel, 38, 44; and survival, 23
negative reciprocity, 76–77
nepotism, 104
Nesse, Randy, 30, 182
Netting, Robert McC., 110
network reciprocity, 75–76
Newark Liberty International Airport, 125–26
New Guinea, 121–22. *See also* Papua New Guinea
Newton, Isaac, 164
Ngenika, 121
niche construction, 28
Nolin, David, 142
nonexcludability, 9
nonhumans: collective behavior of, 154–55, 163; community fissioning in, 41; culture of, 32; emergence in behavior of, 154–55; hierarchies in, 36; human characteristics shared by, 32–34; human psychological characteristics shared by, 33; and novel environments, 29; oxytocin and social attachments in, 73; reciprocal altruism absent in, 77; signaling behaviors of, 91; thought processes of, 34
norms. *See* social coordination norms
North, Douglass, 103, 115
novel environment problems, 29, 43, 182
Nowak, Martin, 7, 75–76, 87
Nozick, Robert, 159
nutrition, 29
Nyamwezi, 171–72

Old Age Revolving Pensions, 63
Old Order Amish, 41
Oliver, Pamela, 7, 195n31
Olson, Mancur, ix, 4–5, 9, 12, 48, 50–52, 56–60, 66–71, 115, 181, 183–85, 194n31; *The Logic of Collective Action*, 4, 45–46, 48, 50–52, 56–59
One Hundred and One Dalmations (movie), 164
ontogenetic explanations, 19
Opp, Karl-Dieter, 66
organizations, 101–23; building on success of other, 104–6; and cooperation, 13;

failures of, 111–15; institutions distinguished from, 103; selection mechanisms among, 115–23; successful, 103–10, 185
Orwell, George, *1984*, 137
osotua (umbilical cord; risk-pooling partners), 118, 145–49
Ostrom, Elinor, 13, 109, 110, 113, 170
overpopulation, 35–37, 47
oxytocin, 73–74

Paciotti, Brian, 172
Page, Scott, 153
pair bonds. *See* long-term pair bonds
Panchanathan, Karthik, 88
Papua New Guinea, 16, 38. *See also* New Guinea
path dependency, 24, 115
patrons, 64–66
Patton, John Q., 122
Paul, Saint, 98
Pavlov strategy, 81
peasant rebellions, 60–61, 64
Penn Station, New York City, 10, 126
Phoenix, Arizona, 169–70
phylogenetic level of explanation, 20
phylogeny, 24–27
physics, 3
Picts, 122–23
Pimbwe, 1, 171–72
Poland, 107
Polanyi, Michael, 154
police, 1, 171–72
political opportunity structures, 106–8
Population Connection, 47, 56. *See also* Zero Population Growth (ZPG)
positive externalities, 22
power law curves, 163–68, 204n54
preference falsification, 166
preference voting, 104, 111
Price, George, 38
Price, Michael, 83–84, 196n29
Price equation, 38–39
priming effects, 202n52
Printing Industries of the Twin Cities, 69, 70
Prisoner's Dilemma, 8, 78–81, 96, 129–30, 132
private goods: excludability of, 54; government provision of, 54–55; public as beneficiary of, 65–66; subtractability of, 53
privileged groups, 49, 56, 64

probability, 163
prosociality, 95, 118–19
proximate explanations, 12, 19–20, 72
pseudoreciprocity, 21
psychology: coalitional, 119–23, 141, 185; cooperation in relation to, 12–13; evolutionary, 90–91, 205n23; evolution of, 28, 33; folk, 90–91; identification of behavioral characteristics, 82–84, 89–90, 185; principles of, in the social sciences, 180–81; and Theory of Mind, 127–28
public good, the, 54
public goods: characteristics of, 9; coercion for provision of, 51–52; common-pool resources compared to, 108; excludability of, 53–54, 108; hunting rewards as, 96–97; obstacles to provision of, 10; the public good vs., 54; subtractability of, 53, 108
Public Goods Games, 17, 41–43, 89, 90, 99, 121public goods problems, 10. *See also* collective action dilemmas
public transcript, 137, 166
punctuated equilibria, 165
punishment: games studying, 15–17; size of society as factor in, 43; walking away as, 82. *See also* Third-Party Punishment Games
purposive benefits, 60–61, 66–67. *See also* expressive benefits
Putnam, Robert, 104–5, 111
Puurtinen, Mikael, 121

Quervain, Dominique de, 20
qwerty keyboards, 161

race, 120
Rapoport, Anatol, 81
rational self-interest: behavior contrary to, 67, 69–70; and collective action, ix, 4, 48, 50–51; emergence and, 203n29; theory based on, 195n46. *See also* self-interest
Rawls, John, 35
Rebel Without a Cause (film), 134
reciprocal altruism, 7–8, 77
reciprocity: altruism and, 7–8, 76, 77; and coalitional psychology, 119; cooperation in relation to, 86; defined, 75; and identification of cooperators, 74, 77–78, 80–84; indirect, 75, 86–92, 197n41; and it-

eration, 78, 80–81; origins of concept, 77–78; pseudo-, 21; reputation and, 75, 86–92; strong, 41–43, 76; uses of term, 75–77
Reese, Frederick D., 61
religion, 98–99
reproduction: evolutionary role of, 23; individual- vs. group-level considerations in, 36–37
reputation: as collective action incentive, 61, 131; language and, 34, 91–92; reciprocity and, 75, 86–92
revolutions, 165–66
Richerson, Peter J., 27, 86, 92, 116, 139, 182, 203n29
Rigdon, Mary, 90
rivalry (goods), 9–10, 13, 53, 108, 170
rivalry (groups), 120, 121
Roatán, 112, 113
Roberts, Gilbert, 95
Robert Wood Taylor Homes, Chicago, 105–6
romantic love, 99
Rothenberg, Lawrence, 66
Rousseau, Jean-Jacques, 131
rowing, 129–30
Rowing Game, 129, 150
rule formulation, 109
Russian Revolution (1917), 166

Sahlins, Marshall, 42, 76–77
Salamon, Sabrina Deutsch, 94
Salisbury, Robert, 61–64, 194n20
Sally-Anne test, 127–28
Sawyer, R. Keith, 153, 161–62
Schelling, Thomas, 10, 13, 97, 124, 126, 136, 160, 204n47
schemata, 145
Schneiderman, Eric T., 125
Schotter, Andrew, 138
science: complementarity of theories in, 5–6, 14, 72, 175–83, 187; coordination problems in, 168; disciplinary relations in, 177–78; historical vs. theoretical, 3; invisible-hand explanations in, 158
sclera, 126–27
Scott, James, 137, 166
scripts, 144–45
sculling, 129–30
Sculling Game, 129
Seabright, Paul, 140
Searle, John R., 102–3

second-order collective action dilemmas, 51, 88–89
Seeley, Thomas, 155
segregation, 160
Seinfeld, Jerry, 119
selection, types of, 22–24. *See also* natural selection
selective benefits, 57–69, 184–85
self-interest, dominance hierarchies and, 36. *See also* rational self-interest
selfish genes, 35, 36
September 11, 2001 terrorist attacks, ix
sexuality, organizational control of, 103
sexual reproduction. *See* reproduction
sexual selection, 23–24, 25
shame, 61
shared attention, 127
shared intentionality, 127
shared knowledge. *See* common knowledge
Shiwiar, 85
Shuar, 83–84
sickle-cell anemia, 33
Sigmund, Karl, 87
signaling theory, 93–99, 155, 177
signals. *See* hard-to-fake signals
Siuai, 157
slave rebellions, 1–2, 172–74
Smith, Adam, 14, 154, 156, 186
smoke detector principle, 30, 182
Snowdrift Game, 22, 134, 150
Sober, Elliot, 39–41
social and behavioral sciences: assumptions underlying, 180; biological group selection promoted in, 38–42; causal explanations in, 12, 20, 158–59; on cooperation, 71; emergence in, 154–56; evolutionary biology in relation to, 11, 14, 72, 175–83, 187; evolutionary explanations in, 44–45; future of, 205n12; human niche construction studied by, 28; invisible-hand explanations in, 158–59; natural vs., 179–80; psychological principles and assumptions underlying, 180–81
social coordination norms, 148, 151–52, 159, 163
social goods, 9. *See also* public goods
social movement organizations (SMOs), 63
social movements, 60, 63, 66, 106–8
social rules, violation of, 84–86
Social Security, 63–64
Social Security Act (1935), 64

social selection, 23; cheater detection, 84–86; cooperative partner choice, 82–84; hard-to-fake signals, 92–99; indirect reciprocity, 86–92
social structure, 162
Soler, Montserrat, 98–99
Solidarity movement, 107
solidary benefits, 60–66, 72, 119, 131, 181
Sopher, Barry, 138
Sosis, Richard, 98–99
Southern Christian Leadership Conference, 132
Soviet Union, 116, 137
species-level benefits of adaptation, 35–36
Spence, Michael, 93–94
spontaneous order, 154
sports teams, identification with, 119–21
stable emergents, 162
staff, of membership groups, 62–63
Stag Hunt Games, 131–32, 138–40, 150, 165
standing, 88
states: coercion and collective action in, 51–52; common-pool resource management by, 109, 170; coordination regulated by, 152; provision of goods by, 54–55
statistics, 163
Steadman, Shannon, 111–15
strategic costs, 93
strategic voting, 133
strikes, 166–67
strong reciprocity, 41–43, 76
substrate neutrality, 177–78
subtractability, 53, 101. *See also* rivalry
Sugden, Robert, 88, 119, 145, 154, 159
Sukuma, 1, 171–72
Sungusungu, 1, 171–72
Super Bowl, 137–38
survival, evolutionary role of, 23
Sutton, Robert, 74, 81
Szathmáry, Eörs, 129

Tanzania, 1, 171–72
Tarrow, Sidney, 106
teams. *See* sports teams, identification with
Tenerife, Spain, 2, 174–75
terreiros (congregations), 98–99
testosterone, 74
theoretical sciences, 3
Theory of Mind, 127–28
third-party candidates, 133

Third-Party Punishment Games, 16–17, 41, 43, 121–22
Thompson, Bill, 125
threshold effects, 165
Thurnwald, Richard, 157
Tillock, Harriett, 47
Tinbergen, Niko, 19
tipping points, 165
Tit-for-Tat strategy, 81–82
Tocqueville, Alexis de, 166
tolerated theft/scrounging, 21
toll goods: excludability of, 54; subtracta-
    bility of, 53
Tomasello, Michael, 127
Tooby, John, 84–85
Törbel, 110, 113–14
Townsend, Francis E., 63
Townsend, Walter, 63
Townsend Movement, 63–65
traffic jams, 160
Tragedy of the Commons, 101, 170
Trivers, Robert, 7, 77, 78, 193n110
trolley car dilemma, 73–74
Truman, David, 48
trust, 132
Trust Game, 16, 147–49
twin studies, 72–73
Tylor, E. B., 31

Uganda, 171
Ullmann-Margalit, Edna, 124–25, 158, 159
ultimate explanations, 12, 19–20
Ultimatum Games, 15–16, 41, 42, 74, 122, 172
Utila, Honduras, 111–15

Valencia, Spain, 170
Van Huyck, John, 138–39
variation, 22
Venkatesh, Sudhir, 105
vigilante movements, 1, 171–72
Visigoths, 122–23

voting: motivations for, 182; preference, 104, 111; strategic, 133
Voting Rights Act (1965), 107, 108
Vrba, Elisabeth, 26

Walk Away strategy, 81–82
Walker, Jack, 64, 67
Walras, Leon, 156
Walston, James, 111
Warsaw Pact, 116
Wasielewski, Helen, 148
Wason selection task, 84–85
water management, 1, 143–44, 169–71
weak emergentism, 177
weak reciprocity, 76
Web of Science, 39, 59
Wedekind, Claus, 88
West, Stuart, 20
whaling, 141–43
wildfires (information), 165
Williams, George C., 4–5, 11, 18, 21, 24, 34–39, 44, 86, 87, 183–84; *Adaptation and Natural Selection*, 4, 18, 34, 37, 38
Wilson, David Sloan, 38–41, 158
Wilson, Edward O., 6, 176, 205n12
Wilson, James Q., 59–61, 72, 73, 119
wings, 25
Win-Stay, Lose-Shift strategy, 81–82
Wolimbka, 121
workplace signaling, 94
Wynne-Edwards, V. C., 36, 37

Yąnomamö, 41
Yamagishi, Toshio, 83
Young, H. Peyton, 159

Zahavi, Amotz, 95
Zero Population Growth (ZPG), 47–48, 50, 52. *See also* Population Connection
Zhong, Chen-Bo, 90
Zipf, George, 164

CPSIA information can be obtained
at www.ICGtesting.com
Printed in the USA
LVHW111119180421
684828LV00006B/1278